Cover: Ocean Sunset. In the Public Domain. The sun is in the early stages of a nova, threatening the lives of billions of people, but God has inspired new inventions to reverse the nova on the sun and rejuvenate and regenerate the earth.

Rejuvenating the Sun and Earth Through God Inspired Science

By Gordon L. Ziegler

Rejuvenating the Sun and Earth Through God Inspired Science

© 2015 Gordon L. Ziegler

Last revised August 11, 2015

Author
Gordon L. Ziegler
P.O. Box 1162
Olympia, WA 98507-1162
ben_ent100@msn.com

PREFACE

When I was a child, the sun was yellow. It was published in child books as being yellow. Except for sunrises and sunsets, that is no longer true today! The sun is now white and very bright. And it is getting brighter and hotter every day. That is because our sun is in the early stages of a nova! This is the principal cause of global warming. Carbon dioxide and greenhouse gases are part of global warming, but not the principal part. Not only is the earth's polar cap melting, but the polar cap on mars is melting also! That is not caused by greenhouse gas emissions from coal power plants on earth. That is caused by a warming on the sun. Humans can do something about Carbon emissions, but it has been thought that humanity can do nothing about a nova on the sun! But that is no longer true! There **is** something now that humanity can do to reverse a nova on the sun! God has inspired a science that can reverse the nova on the sun as well as rejuvenate or regenerate the earth!

What would happen to earthlings if we did nothing to the sun? The sun would get brighter and brighter more and more quickly until the sun would give the light of seven days in a single day, and the moon would become as bright as our sun is now! Isaiah 30:26. The sun would "scorch men with fire. . .with great heat!" Revelation 16:8 and 9. Then, if the nova sequence on the sun continued non-mitigated, the sun would decrease and decrease in brightness and go out in darkness, deep freezing the earth! But the Bible predicts that God will provide that men will not even need the light and heat of the sun! Revelation 21:23. All the glory and honor will go to God and the Lamb Jesus, but there is something men can do to cooperate with God now to save earth from a solar holocaust and great sufferings and a wave of deaths.

What causes a nova on the sun? A new star is formed through self gravitational attraction of a large cloud of Hydrogen gas. The gas cloud implodes enough to have a thermo-nuclear ignition of the Hydrogen fusing to Helium in the core of the sun, releasing light and heat enough to stop the implosion of the gases, and shine as a sun or star. This reaction lasts quite awhile, and the sun or star shines a long time—fusing Hydrogen to Helium and shining. But when the sun or star becomes very old there is a phase change in the core of the sun when the Helium becomes about 50 percent of the Hydrogen-Helium gas mix in the sun. The Helium is more opaque to the light and heat radiated from the sun or star than Hydrogen is. It becomes like a thermal blanket holding in the heat in the core of the star. The core of the star becomes hotter which increases the rate of the hydrogen fusion in the core of the sun. This makes the sun or star hotter and brighter and makes the Hydrogen fuse even faster. But this is depleting the available Hydrogen to fuse. The explosion of light and heat in the core of the sun peaks, then there is no longer enough available Hydrogen for the Hydrogen to fuse at the previous rate, and the core cools down and slows down the Hydrogen fusing rate in the core of the star more and more until the sun or star goes out in darkness, deep freezing the planet or earth orbiting the star or sun. That's a nova! We are in the beginning stages of a nova on the sun!

Scientists have calculated that we have about 500 years before the crisis. But they are wrong! It could be a year or less before we reach the peak. But we haven't reached it yet, though temperatures have exceeded 120 degrees Fahrenheit in several parts of the world (45 degrees Celsius) and thousands of people in India have died from the heat. It appears that we still have a chance to build the Refresher and use it to prevent a temperature peak on the sun as well as regenerate the earth! If Electrino Group Inc obtains a $100,000,000 guaranteed loan from DOE Loans Programs Office, it now appears that we will only need $30,000,000 more (including $3.05 million for loan fees). That is probably more doable. Our current effort is to reach investors with these inventions and other leaders in the clean energy field, to try to get the ball rolling.

CONTENTS

Chapter   Page

Preface...4
Contents..6

1. Reversing the Sun..8
2. The Second Law of Thermodynamics.......................................9
3. The Theory..17
4. Is it Safe? 27
5. What Would the Refresher Do? 36
Reverse aging for adults...36
Resurrections from the dead..36
Backing diseases out of existence..36
Reversing all decay..36
Reversing pollution out of existence.....................................37
"Raising up the foundations of many generations"
Isaiah 58:12...37
Reversing forest fires 37
Reversing all calamities...37
Reversing all effects of war...37
Preventing all munitions from firing.....................................37
"Making wars to cease to the end of the earth." Psalm 46:9..37
Removing sinful propensities from people, including
criminals..37
Emptying prison houses...37
Making possible and efficient Clean Energy Sources.......37
Taming the Sun 37
The blessings of the Refresher are endless...............................37
6. Should We Do It? 38
7. The Messiah's Inventions..43
8. Correcting the Standard Model of Physics..............................45
9. Clean Energy Sources..57
10. Clean Energy Theory..58

11. Project Description..71
A. Introduction..71
B. Sketches...71
C. Costs..74
D. Emergency Situation...75

Chapter 1

Reversing the Sun

We could reverse the nova on the sun if we could reverse the second law of thermodynamics on the sun. It is not as difficult as it sounds. First we need to realize that the second law of thermodynamics on the sun is just the same as we are now experiencing it on earth. The order to disorder arrow of time in the second law of thermodynamics on the sun and earth now points from order to disorder. That is because the order energy on the sun and earth is now slightly negative. If we could make it slightly positive on the sun and earth, we could reverse the order to disorder arrow of time on the sun and earth, and therefore reverse the direction of processes on the sun and earth, on the sun making the Helium fission to Hydrogen again, backing up the Hydrogen to Helium processes on the sun. We could back up the sun several thousand years to safer times, before letting the sun go back to its Hydrogen to Helium fusion processes. We first need a basic understanding of the second law of thermodynamics and Order Energy E_O and Entropy Energy E_S.

Chapter 2

The Second Law of Thermodynamics

A. Introduction

Everything goes from a state of order to more disorder. Brand new automobiles wear out and rust. Objects break or are damaged. A thermos bottle falls off the counter, and the inner glass bottle is shattered. We do not expect the shattered bottle to fall back up to the counter and become whole again. There is a one-way arrow for the events to transpire. That arrow is the order to disorder arrow of time in the second law of thermodynamics.

Houses grow old and fall into decay. Barns fall down. Fruit spoils, people and animals grow old and die. Viruses mutate. People become ill and die. Crime and disorder in society increase. Homes break up. Aborted fetuses disintegrate. Dead people and things decompose. All of these negative occurrences are the outworking of the second law of thermodynamics—that part of which is an arrow making everything go from order to disorder.

Let us consider what other people have written about the second law of thermodynamics.

"Second law of thermodynamics
"An equilibrium macrostate of a system can be characterized by a quantity S (called *entropy*) which has the following properties:

"(i) In any infinitesimal quasi-static process in which the system absorbs heat dQ, its entropy changes by an amount

$$dS = \frac{dQ}{T} \qquad (2\text{-}1)$$

where T is a parameter characteristic of the macrostate of the system and is called its *absolute temperature*.

9

"(ii) In any process in which a thermally isolated system changes from one macrostate to another, its entropy tends to increase, i.e.,

$$\Delta S \geq 0. \qquad (2\text{-}2)$$

"The relation (2-1) is important because it allows one to determine entropy *differences* by measurements of absorbed heat and because it serves to characterize the absolute temperature T of a system. The relation (2-2) is significant because it specifies the direction in which non-equilibrium situations tend to proceed."[1]

The above expression of the second law of thermodynamics is regarding entropy and heat. Other writers include the order to disorder arrow in the second law of thermodynamics.

"It is a matter of common experience that disorder will tend to increase if things are left to themselves. (One has only to stop making repairs around the house to see that!) One can create order out of disorder (for example, one can paint the house), but that requires expenditure of effort or energy and so decreases the amount of ordered energy available.

"A precise statement of this idea is known as the second law of thermodynamics. It states that the entropy of an isolated system always increases, and that when two systems are joined together, the entropy of the combined system is greater than the sum of the entropies of the individual systems. For example, consider a system of gas molecules in a box. The higher the temperature of the gas, the faster the molecules move, and so the more frequently and harder they collide with the walls of the box and the greater the outward pressure they exert on the walls. Suppose that initially the molecules are all confined to the left-hand side of the box by a partition. If the partition is then removed, the molecules will tend to spread out and occupy both halves of the box. At some later time they could, by chance, all be in the right half or back in the left half, but it is overwhelmingly more probable that there will be roughly equal numbers in the two halves. Such a state is less ordered, or more disordered, than the original state in which all the molecules were in one half. One therefore says that the entropy of the gas has gone up. Similarly, suppose one starts with two boxes, one containing oxygen molecules and the other containing nitrogen molecules. If one joins the boxes together and removes the intervening wall, the oxygen and nitrogen molecules will start to mix. At a later time the most probable state would be a fairly uniform mixture of oxygen and nitrogen molecules throughout the two boxes. This state would be less ordered, and hence have more entropy, than the initial state of two separate boxes."[2]

"The explanation that is usually given as to why we don't see broken cups gathering themselves together off the floor and jumping back onto the table is that it is forbidden by the second law of thermodynamics. This says that in any closed system disorder, or entropy, always increases with time. In other words, it is a form of Murphy's Law: Things always tend to go wrong! An intact cup on the table is a state of high order, but a broken cup on the floor is a disordered state. One can go readily from the cup on the table in the past to the broken cup on the floor in the future, but not the other way round.

"The increase of disorder or entropy with time is one example of what is called an arrow of time, something that distinguishes the past from the future, giving a direction to time."[3]

B.   Electrino Model and 2nd Law

The natural tendency of leptons in beta decay is that the parent lepton combines with one or more gravitons to produce more particles. In all natural reactions, the order energy of the resultant particles is less than or equal to the order energy of the original particles.

1.   Negative Energies. Let us consider antimatter more carefully. "In the Dirac theory also, *the permissible energy values for a free particle range from $+mc^2$ to $+ \infty$ and from $-mc^2$ to $- \infty$.* The first of these results is of course just what we expect for a free particle—that its total energy can have any value greater than its rest energy. But the second result

is quite puzzling, since it implies the existence of states of *negative total energy.*"[4] Anderson in 1932 discovered positrons in cosmic radiation. These were regarded as Dirac's negative energy particles. "The first two solutions of the Dirac equation . . . clearly describe a free electron of energy E and momentum **p**. The two negative energy electron solutions . . . are to be associated with the antiparticle, the positron."[5]

However, in the annihilation it is not $(+mc^2) + (-mc^2) = 0$, but $2mc^2$ is the result of annihilation.[6] There is something strange going on with the minus signs in these equations. The calculations are inconsistent.

Maybe there are two kinds of energy considered. One we can call entropy energy E_S. In the annihilation reaction, $|+mc^2| + |-mc^2| = 2mc^2$. Entropy energy is the higher value. The other energy is order energy E_O. In order energy the same reaction is $(+mc^2) + (-mc^2) = 0$.

Let us consider entropy energy and order energy for particle decay schemes. There are a few decay schemes where no negative order energy (anti-matter) is introduced in the right hand side of the decay schemes. In those few instances, the final order energy is equal to the initial order energy (when kinetic energy is taken into account). But in most cases, a trace of negative order energy (anti-matter) is introduced into the right side of the decay schemes. There is nothing on the left hand sides of the decay schemes to correspond to this addition of a trace of negative order energy on the right sides of the decay schemes. Therefore, total order energy is less on the right hand sides of the decay schemes than on the left hand sides (if only by a trace). A few decay schemes introduce a lot of antimatter (as K⁻) on the right side of the decay scheme. The loss of order energy in the systems is greater in those cases. But in every case, for all natural processes, the order energy final is less than or equal to the order energy initial, or

$$\Delta E_0 \leq 0. \tag{2-3}$$

Let us check the order energy for electron electrino fusion reactions. Electrons made energetic by acceleration (as heavy as protons) fuse and form anti-protons. Matter is converted to anti-matter. Entropy energy is conserved, but not so order energy. Order energy is reduced in the extreme from +938 MeV to -938 MeV or more for each electron fused (two electrons are fused in each reaction). The order-disorder arrow for electron electrino fusion points in the usual direction. The system does obey the second law of thermodynamics as we now know it.

2.   Reversing the Order to Disorder Arrow.   What would happen if we fused the electrino constituents of positrons instead of the electrino constituents of electrons? Entropy energy E_S would again be conserved.   Entropy would be increased. However, order energy E_O would go from -2 x 938 MeV to +2 x 938 MeV—from disorder to order. The order to disorder arrow would be reversed. This would be a reaction that would be prohibited by the second law of thermodynamics—unless the strong gravitational force that fuses the anti-semions would be stronger than the second law of thermodynamics (which otherwise governs weak interactions), which it is.

Here we see that the entropy arrow of time and the order to disorder arrow of time are separate and distinct, and are not one and the same thing. While all the reactions the author has studied increase entropy, the fusion of positron anti-semions reverses the order to disorder arrow, making more order out of the disorder.

Positron constituent electrino fusion might not only take the electrinos from disorder to order. It could make other physical processes in a local area go from disorder to order. The positron fusion not only violates the second law of thermodynamics, it reverses the order to disorder arrow of that law in a local area, making other processes in that area reverse. Let us consider that process more to see how it might be regulated.

We guess the desired relationships for reversing the order to disorder arrow in the second law of thermodynamics through dimensional analysis. We want to solve for r, the maximum radius in which the reversed law would be effective. There is a way we can obtain a length from combinations of our variables and constants. That way is in the right hand side of Eq. (2-4). The whole expression is the thermodynamic relation we are seeking. The thermodynamic relation is:

$$(\Delta E_o)_t > 0 \ where \ r < \frac{(\Delta E_o)_1 \ c}{ik}, \qquad (2\text{-}4)$$

where E_o is the order energy–the positive or negative energy in the pair production of particles; ΔE_o is the change in the order energy, where $(\Delta E_o)_t$ is the change in the total order energy of the system, and where $\mathbf{(\Delta E_o)_1}$ is the change in the order energy for a single source reaction—for a positron fusion reaction it is approximately $2 \times 0.94 \times 10^9$ eV/collision $\times 1.6 \times 10^{-19}$ joules/eV = $\mathbf{3.0 \times 10^{-10}}$ $\mathbf{joules/collision}$; $\mathbf{c}$ is the speed of light—approximately $\mathbf{3.0 \times 10^8}$ $\mathbf{m/s}$; we shall solve for the effective radius r; $\mathbf{i}$ is the effective beam collision current in each beam in Coulombs per second (we will solve for $\mathbf{10^{-11}}$ or 10 picoAmps); $\mathbf{k}$ is the ratio of particle energy to particle charge. This energy per charge is the accelerated energy of the particle (0.94×10^9 ev times 1.6×10^{-19} joules/ev = 1.5×10^{-10} joules) divided by the charge of each positron ($q = 1.6 \times 10^{-19}$ coulombs), which equals $\mathbf{9.38 \times 10^8 \ joules \ per \ coulomb}$. The collision efficiency eff is not needed in this equation, because the result is not in particles, but is already in collisions.

Incredibly, the lower the collision rate, the bigger the radius of the affected area. And the greater the collision rate, the smaller the radius of the affected area. With 10^{-11} A effective beam currents (at 100% efficiency), the effective radius $\mathbf{r}$ solves for $\mathbf{9.6}$ $\mathbf{meters}$—which describes a small area—less than a tenth of an acre. Instead of beam current collisions at 100% efficiency, we calculate the net collision rate needed, which could be achievable with much lower particle collision efficiency.

To get an idea of the positron collisions needed to reverse the order to disorder arrow of the second law of thermodynamics in what size of affected radius, see Table 2-1 below.

For footprint the radius of	r	effective collision current	net collisions/sec
house	9.6 m	10 pA	6.2E7
4 ftball flds	96 m	1 pA	6.2E6
community	960 m	100 fA	6.2E5
city	9.6 km	10 fA	6.2E4
Israel	160 km	0.6 fA	3,800
U.S.	2,400 km	0.04 fA	250
World	13,000 km	0.008 fA	46
Sun	1.7E11 m	6E-22 A	1 collision/79 hrs

Table 2-1. Net collision rate and effective collision current of positrons at 940 MeV colliding with spin flipped positrons at 940 MeV necessary to make a Refresher footprint of given radius for reversal of the order to disorder arrowof the second law of thermodynamics. [To calculate the net collisions/sec, multiply the net equivalent collision currents by 6.25E18.]

Remarkably enough, the affected area of second law reversal calculates to increase with the reduction of positron effective beam current. Area control is merely a matter of timed gating of the positrons in the positron-positron collider.

[1] F. Reif, *Statistical Physics*, Berkeley Physics Course— Volume 5 (New York: McGraw-Hill Book Company, 1967), p. 283.

[2] Stephen Hawking, *A Brief History of Time*—From the Big Bang to Black Holes (New York: Bantam Books, 1988), pp. 102, 103.

[3] *Ibid.*, pp. 144, 145.

[4] Robert B. Leighton, *Principles of Modern Physics* (New York: McGraw-Hill Book Company, Inc, 1959), p. 665.

[5]Francis Halzen, Alan D. Martin, *Quarks and Leptons* (New York: John Wiley & Sons, 1984), p. 107.

[6]David S. Saxon, *Elementary Quantum Mechanics* (San Francisco: Holden-Day, 1968), p. 386.

Problem Set 2

1.   "Humpty-Dumpty sat on a wall. Humpty-Dumpty had a great fall. And all the king's horses and all the king's men couldn't put Humpty-Dumpty back together again." What law of physics does this child's nursery rhyme illustrate? What arrow of time is demonstrated?

2.   Is the entropy more or less after two gas containers are opened to each other?

3.   You watch a movie. Broken pieces of glass fly up and become a window. Is the movie playing forward or backward? How do you know?

4.   What generally happens in beta decay? Is the system going to more or less order?

5.   What would the world be like if the order to disorder arrow were reversed?

Chapter 3

The Theory

Refresher

The Principal Investigator has discovered a new Grand Unification Theory (GUT). It has deeper symmetry and lower orbital structures than the Standard Model of Physics. It has greater parsimony than the Standard Model. Whereas the Standard Model requires 61 different elementary particles to construct known light and matter [1](page 48), the GUT requires only two different elementary particles to construct known light, matter, and gravitons; and those two different particles can both be ionized from empty space with a single particle, and can combine in a single particle.

What differences does this GUT have to the Standard Model? The GUT is an aether model of physics. It has aether special and general relativity, rather than Einstein's aether-less Special and General Relativity. This makes a simple model of gravity and inertia possible [2](Chapter 5). Up until now, uniting special and general relativity in particle physics has been as difficult as uniting fire and ice. This problem is solved with aether special and general relativity in the GUT [2](Chapter 6). Special and general relativistic calculations are both exact fits in the particle structures calculated in that chapter.

The GUT has one postulate that states that symmetric smooth charge distributions cannot have detectable spin. But electrons and positrons have detectable spins. Therefore they must not be symmetric point charges, but have two half charges in them orbiting about each other. The orbiting like charges show that fracton charges come in $\pm e$, $\pm e/2$, $\pm e/4$, and $\pm e/8$ (the Electrino Hypothesis), rather than in $\pm e/3$ and $\pm 2e/3$ (the Quark Hypothesis). The Electrino Hypothesis is very different from the historic and accepted Quark Hypothesis. Yet it does not lead to untenable particle structures. The Principal Investigator has induced the particle structures of all known light, matter, and gravitons, through the simultaneous satisfaction of ten criteria: particle charge, spin, parity, mass, spin feasibility, preceding particles (to avoid duplication), the Pauli Exclusion Principle, b state laws, decay schemes, and the requirement that no particles except electrons and positrons and neutrinos and anti-neutrinos have ground state – and + echons in them, so no particles other than electrons and positrons have to have defined masses in them. [2](Appendix B). Satisfying all the listed Decay Modes published in *Summary Tables of Particle Properties*, by the Particle Data Group [3] (which reference [2](Appendix B) does), is the satisfaction of thousands of tests. Except for being an unknown model, the GUT is in a strong position. Its particle structures are all unique. The quark model particle structures are not all unique. The quark model has seven formulas for eight particles. The first and the last are the same. But the masses of the first particle and the last particle are not the same! The quark model fails.

Another difference of the GUT to the Standard Model is that charged sub-particles of like charge fractions can fuse to particles of higher charge fractions. [2](Chapter 12). The secret of why that should be is that when sub-particles orbit or travel faster than the speed of light in the relativistic frame relative to the baseline non-relativistic frame, their radii become imaginary because of the relativistic length contraction formula. The strong electric force equation for these super luminal sub-particles has two such imaginary radii multiplied together. That makes an additional minus sign in the force equation, which makes like charges attract. When two bound sub-particles of a positron collide in the Center of Mass Frame with two other bound sub-particles of another positron with like oriented spins in the Center of Mass Frame, with 1880 MeV energy or more, the four sub-particles are attracted into the same orbit. Then one sub-particle from one positron is more attracted to one sub-particle from the other positron than to any other sub-particle because of closer proximity. These are like charges, and here like charges attract because they travel faster than light. [Einstein predicted that nothing could go faster than the speed of light. But neutrinos have recently been clocked as going faster than the speed of light. It appears that Einstein was wrong. [4-6]] The two sub-particles are attracted by the electric strong force. Nothing stops them from fusing. The other two positron sub-particles fuse also. Four sub-particles fuse down to two particles each with twice the charge as the charge of one of the four sub-particles. The four sub-particles are ½ e charges. The two fused particles are 1 e charges each, but though they are numerically whole particles, they cannot exist alone. That is why on creation they scavenge from the graviton sea the necessary sub-particles to become protons or neutrons. But when the positron four ½ e sub-particles fuse to the two 1 e particles, they switch from antimatter to matter. The fusion of sub-particles in positrons results in the generation of solely positive order energy (quantum mechanical energy in the creation of particles). This phenomenon is theorized to reverse the order to disorder arrow in the second law of thermodynamics [because it is positive order energy as opposed to negative order energy which surrounds us and which determines the current order to disorder arrow direction and the direction of reactions]. [2](Chapter 16).

The fusion of the sub-particles of positrons can result in the reversal of the order to disorder arrow in the second law of thermodynamics—but over what distance? Those answers are already derived in Chapter 2 (pages 13-15).

Incredibly, the lower the effective current (or the lower the collisions per second) the bigger the radius of the affected area. And the greater the effective current, the smaller the radius of the affected area. With 10^{-11} A effective collision currents, the effected radius r solves for 9.6 meters, which describes a small area—less than a tenth of an acre.

To get an idea of the positron net collision currents needed to reverse the order to disorder arrow of the second law of thermodynamics in what size of affected radius, see Table 2-1 in the last chapter.

We don't need this in the Clean Energy Source, but the Refresher needs electronics to gate the colliding beams by at least thirteen decades with a control board key-lock position for each of switch settings decades below 100 pA/eff net collision rates. (To obtain the collisions per second, multiply the effective collision currents by 6.25E18.) Initially we will not know the efficiency eff. We will not be correct, but until we can measure the efficiency of collisions, we can assume it is 1.0. You can use as much beam power as you want up to 1.0 Amp to achieve initial collisions, but then gate it down to 100 pA; 10 pA 1.0 pA; 0.1 pA; 10 fA; 1.0 fA; 0.1 fA; 0.01 fA; 0.001 fA; 1.0E-19 A; 1.0E-20 A; 1.0E-21 A; and 1.0E-22 A. The gating needs to be before the beams collide. The spin flipper on one of the beams appears to be the best place to gate the beams: If you don't flip the beam for a time, or you flip it less than 90 degrees, the beam will shoot blanks, or essentially the beam will be chopped for a time. For reference sake let beam current height times total beam current width be 1.0 pA/eff for beam collisions for on the order of 100 meters radius foot print. Let lower net beam currents be same current heights and peak widths, but with longer times between gated peaks. This will increase footprint radius. (It is the inverse of what you would think. It is counter intuitive.) It is essentially an electronic clock hooked to the beam flipper.

The author will now calculate the rate at which reverse aging will occur in the calculable radius of the active Refresher: The beginning energy of the particles (positrons) from which the fusion process takes place is $2m_ec^2$ per individual reaction. The ending energy of the particles (protons) to which the fusion process tends is $2m_pc^2$ per individual reaction. $\dfrac{\Delta E_p}{\Delta E_{e^+}} = \dfrac{+2m_pc^2}{-2m_ec^2} \approx -1836.$

This is a unit less expression from the available energy terms. What we seek is another unit less expression $\dfrac{\Delta t_r}{\Delta t}$, where t is the normal time during which a person or object ages, and t_r is the reverse time (negative) during which a person or object un-ages. The quotient is the relative rate of un-aging compared to aging. This also is a unit less quotient. What use of particle fusion parameters can yield such a unit less quotient? What terms are available to derive such a unit less quotient? What about the first terms and unit less quotient? If we equate them, we have

20

$\dfrac{\Delta t_r}{\Delta t} \approx -1836$. Reverse time would be negative and 1836 times as fast as forward aging time. Forward aging of 100 years would be un-aged in 19.89 days. Forward aging of 1 year would be un-aged in approximately 4.77 hours of machine time.

The Refresher would have many different effects in a controlled area, and the controlled area could be varied in size from the thickness of pencil lead or less to the earth orbit of the sun. Some such effects would be reverse aging; backing diseases out of existence; backing decay and pollution out of existence; disaster and war relief; making a demilitarized zone in the controlled area where no explosives will explode; building and vehicle maintenance; making a new form of food preservation; removing criminal tendencies in brains; taming animals; and making Clean Energy Sources possible; etc.

Placing a Clean Energy Source in the footprint of the Refresher would prevent the creation of radioactive wastes in the Clean Energy Source, because the force of the positive sense disorder to order arrow in the Refresher would be about 1836 times as powerful as the negative sense order to disorder arrow in the Clean Energy Source. The power of the Refresher disorder to order arrow would over power the force of the Clean Energy Source order to disorder arrow.

But would the power of the Refresher disorder to order arrow of time prevent the sub-particles of electrons in the Clean Energy Source from fusing? Thank God the answer is no! The gravitational and electrical strong forces attracting the semions in electrons in the Clean Energy Source are the strongest forces of all—stronger than the positive sense disorder to order arrow in the Refresher! Where we want radioactive wastes to be suppressed in the Clean Energy Source, they are suppressed. And where we want the fusion processes not to be suppressed, they are not suppressed! The Refresher and the Clean Energy Source appear to be made for each other—a match made in heaven!

1. How can scattered glass fragments be re-gathered together and fused into a perfect window without the actual backing up of the clock? The order to disorder

arrow in the second law of thermodynamics is a powerful force: it is more powerful than gravity; it is more powerful than the force of a speeding projectile; it is more powerful than chemical bonds; it is more powerful than surface tension or friction; it is more powerful than atomic or hydrogen bombs! It is so powerful that nothing that science could do could resist it—up until now! For thousands of years the whole human race has been controlled and defeated by the order to disorder arrow of time turned in the negative sense. Thank God, there is something in nature that is more powerful than the order to disorder arrow in the second law of thermodynamics! It is the strong electric and strong gravitational forces. But their action is confined in extremely small spheres. How could they be controlled and made to work in man's benefit? The aether is the key here! Einstein's aether-less relativities will not work here! The fusion of electrinos—anti-semions into unitons in those extremely small spheres is accomplished by those resistless and non-reversible strong forces. The process produces positive order energy and the creation of new protons. The aether penetrating those extremely small spheres pass through and somehow carry away the information that new protons are being created in this vicinity. That information carried by the outgoing aether flips the order to disorder arrow to the positive sense, which now is a super powerful force to re-create everything! An array of broken glass fragments has an order to disorder value. In the active area of the Refresher or Regenerator machine [for which we have already calculated its effective radius based on effective collision currents], the system is forced in a small increment of time to move the glass shards to a slightly more positive order to disorder value. It keeps doing this one Δt and one ΔE_O at a time until the glass shards fall back up and fuse into the original window. This type of process will work for re-uniting severed spinal cords and healing and restoring war wounds.

2. Cystic fibrosis and Down's syndrome victims are cured by another process of this order to disorder reversal. The DNA code that God made for every living thing is a powerful self-correcting thing from most mutations. But not in the cases of Cystic fibroses and Down's syndrome victims in the current negative order to disorder arrow sense. But when the order to disorder arrow is flipped to the positive sense by the creation of new protons, there is a powerful force in all the bodies to restore the DNA and RNA codes in every cell to the maximum order correct codes. Every mutation of the DNA and RNA codes is deleterious, as was demonstrated by the irradiation of fish egg experiments in the Fisheries Department of the University of Washington USA in the 1970s. There are gillions of wrong code values, but only one maximum order code for every conception code. When the order to disorder arrow in the second law of thermodynamics is flipped to the positive sense, there is a powerful force in every cell of the body to right the existing codes to the correct codes. That is why the DNA and RNA codes in every cell are righted to the original divinely desired code in each cell to the same code. This heals Cystic fibrosis and Down's syndrome cases as well as not desired skin colors and birth deformities. Everyone's codes are restored to the correct divinely intended conception codes.

3. Every disease germ or cell or virus has negative code values. The good pro-biotic bacteria all have positive codes. Reversing the order to disorder arrow in people and their environs backs the harmful bacteria and viruses out of existence without harming the good bacteria as antibiotics do. This can back almost every

disease out of existence including HIV AIDs, cancer, and Ebola.

But how can we penetrate those extremely small spheres confining the strong forces, to induce an anti-semion to uniton fusion reaction? Normally electron or positron beams have a 1.602E-19 efficiency of colliding. But microscopic fine beams even at that efficiency can have one or two or a few collisions. This collision efficiency works for electrons as well as positrons; but for electrons it results in a negative sense for the order to disorder arrow, which can be reversed only by a stronger positive sense of the order to disorder arrow from the fusion of positron anti-semions to unitons, the core particles of protons and neutrons. But fortunately the positive sense of the order to disorder arrow in the second law of thermodynamics from the fusion of positron anti-semions to unitons is about 1836 times as powerful as the negative sense of the order to disorder arrow of time in the second law of thermodynamics by the fusion of electron semions to anti-unitons. This permits Clean Energy Sources to work in the presence of a Refresher. But back to the penetration of the extremely small spheres by positrons: it just takes one collision above 1876 MeV in the center of mass frame [with axial spins, with the spins from one positron accelerator flipped relative to that accelerator, so that the semions from both accelerators have the same sense on colliding] to send the aether outward in a diverging rays road map focusing all successive positrons to that location of proton creation. The diverging aether rays make a three dimension and time roadmap not only in the outer non-relativistic frame, but also in the inner relativistic imaginary dimensions into the extremely small [on the order of E-43 imaginary meters] spheres containing the strong forces. Just one anti-semion fusion event [four semions to two unitons] creates a special roadmap for successive positron anti-semion replications. But remember that just one fusion event will affect a much larger volume in reversing the order arrow, than a few or many fusion events. The efficiency of the events is just the inverse of what we expect. The system is counter intuitive. On starting the chain of events, there would almost certainly be a solar system wide positive order energy transient, but it would not necessarily be long lasting.

When we stop to ponder the restorative processes, they are less like unbelievable miracles, and more like desirable understood processes. Reversing forest fires would be the first type of process on a grand scale. Conceptually it is simple rapid Δt, ΔE_O movements of particles and gases, recombining them according to their original codes.

Do these added insights into electrino fusion restorative processes help any in relieving disbelief in the seemingly miraculous impossibilities?

The Refresher would be quite versatile: It could be built on land, in or on ships, in large airplanes, in long rail cars, in long semis, and in space craft and space bases. It would revolutionize the earth and would bring peace to earth.

---

[1] David Griffiths, *Introduction to Elementary Particles* *(New York: John Wiley &Sons, Inc., 1987)*.

[2] Gordon L. Ziegler, *Electrino Physics* (PO Box 1162, Olympia, WA 98507-1162 USA: Benevolent Enterprises) [downloadable for free at http://www.benevolententerprises.org Book List]. *Electrino Physics* Draft 2 is now available at createspace.com, amazon.com, and barnsandnoble.com for less money.

[3] C. Caso, *et. al.* (Particle Data Group), "Summary Tables of Particle Properties," (including "Gauge and Higgs Bosons Summary Table," "Lepton Summary Table," "Meson Summary Table," and "Baryon Summary Table), *CRC Handbook of Chemistry and Physics*, 80th Edition, David R. Lide, Ph.D., Editor-in-Chief (Boca Raton: CRC Press, 1999-2000), pp. **11**-1 to **11**-42.

[4] Frank Jordans and Seth Borenstein, "Neutrinos clocked moving at faster-than-light speed" (Associated Press):

nce-science/?gt1=43001

[5] Robert Evans, "Particles found to break speed of light"
(Reuters):
http://www.reuters.com/article/2011/09/22/us-science-light-
idUSTRE78L4FH20110922

[6] "Do neutrinos move faster than the speed of light?" –
physicsworld.com:
http://physicsworld.com/cws/article/news/47283

Chapter 4

IS IT SAFE?

Up to this point, *Electrino Physics* has deliberately been without reference to God or the Bible. This has been so as not to prejudice the non-believing scientist. However, in this chapter we come to an urgent question of science for which current science has no data. We need to know "is the reversing of the order to disorder arrow in the second law of thermodynamics safe?" While science has no data on this, it is interesting to note that the Bible and the writings of a modern prophet, Ellen G. White (1827-1915), have rather a lot to say about the phenomena of interest.

The author, in over fifty years of research in the Bible and the writings of Ellen G. White, has found them both to be authoritative, comprehensive, reliable, and 100 percent accurate. So far, that has not been true of any other ancient or modern prophet, prognosticator, fortune teller, astrologer, sorcerer, or soothsayer. In this author's opinion, the Bible writers and Ellen G. White were inspired by God. That is why their statements are so accurate. The author will cite both the Bible and Ellen G. White in this scientific chapter. The non-believing scientist may consider the reasonableness of the material in this chapter, and form his or her own opinions.

Today we live with the second law of thermodynamics and Murphy's Law where everything tends to go wrong, order tends to disorder. The described phenomena after the moral fall of Adam and Eve in Genesis Chapter 3 and Ellen G. White's writings are just the same as we observe today. Yet Adam and Eve previously existed in a different state—a different world order—before their fall—a world order that knew no aging, disease, decay, or death. In our study of the second law of thermodynamics in the last chapter, we have found only two states to the second law of thermodynamics—the current state, and a reversed state. Apparently only two states of that law exist. Then, if Adam and Eve experienced two states in their lifetimes, and the latter state is just what we experience now, then the original state must be the reversed state of the order to disorder arrow of the second law of thermodynamics. The safety information we seek of the reversed state of the order to disorder arrow of the second law of thermodynamics, then, is just the relative safety of the state of being before the fall of Adam and Eve compared to the relative safety of the state of being after their fall.

We can learn a lot about the original state by recognizing the differences that were noted immediately after the fall:

After his transgression Adam at first imagined himself entering upon a higher state of existence. But soon the thought of his sin filled him with terror. The air, which had hitherto been of a mild and uniform temperature, seemed to chill the guilty pair. The love and peace which had been theirs was gone, and in its place they felt a sense of sin, a dread of the future, a nakedness of soul. The robe of light which had enshrouded them, now disappeared, and to supply its place they endeavored to fashion for themselves a covering; for they could not, while unclothed, meet the eye of God and holy angels.[1]

Let us chart the phenomena mentioned in the above paragraph:

States of the Order to Disorder Arrow
Of the Second Law of Thermodynamics

Phenomenon	Post-Fall State	Pre-Fall State
Air Temperature	Chilly	Mild, Uniform
Mental State	Guilt, Fear	Love, Peace
Natural Covering	None-Nakedness	Robe of Light

Let us continue to chart the contrast between the pre-fall and the post-fall states of Adam and Eve and their surroundings.[2,3]

28

States of the Order to Disorder Arrow
Of the Second Law of Thermodynamics

Phenomenon	Post-Fall State	Pre-Fall State
Attitude at Approach of Creator	Fled in Terror	Joy, Welcome
Facing Sins, Mistakes	Self Justification Casting Blame	
Lot of Women	Sorrow, Pain, Multiplied Conception	No Sorrow or Pain
Relationship to Husband	Subjection to	Equal to
Ground	Cursed, Bring Forth Thorns And Thistles	No Curse, No Thorns, No Thistles
Labor Lot of Race	Anxiety, Toil Disappointment, Grief, Pain, Finally Death	Happy Labor No Grief, Pain, or Death
Inferior Creatures	In Revolt Against Man	Subject to Man
All Life on Planet	Mortal	Conditionally Immortal

As we compare the post-fall state with the pre-fall state, we note in all cases that the post-fall state is not the safer state. It is the dangerous one. In every way the pre-fall state is the safer state. The post-fall state is just the condition we find ourselves in today. And we are subject to the second law of thermodynamics. In our study of that law, we have found only two states of that law. And our current state of the second law of thermodynamics is just the same as the post-fall state. Therefore the pre-fall state must be the second state of the second law of thermodynamics—the reversed state we are studying. The study of the experience of our first parents is an evidence of the safety of the reversed state of the second law of thermodynamics.

Bible prophecies of heaven and the earth made new are other evidences of the safety of the reversed state of the second law of thermodynamics. "There, 'the wilderness and the solitary place shall be glad for them; and the desert shall rejoice, and blossom as the rose.' 'Instead of the thorn shall come up the fir tree, and instead of the brier shall come up the myrtle tree.' 'The wolf also shall dwell with the lamb, and the leopard shall lie down with the kid; . . . and a little child shall lead them.' 'They shall not hurt nor destroy in all My holy mountain,' saith the Lord. Isaiah 35:1; 55:13; 11:6, 9.

"Pain cannot exist in the atmosphere of heaven. There will be no more tears, no funeral trains, no badges of mourning. 'There shall be no more death, neither sorrow, nor crying: . . . for the former things are passed away.' "The inhabitant shall not say, I am sick: the people that dwell therein shall be forgiven their iniquity.' Revelation 21:4; Isaiah 33:24. . . .

"In the City of God 'there shall be no night.' None will need or desire repose. There will be no weariness in doing the will of God and offering praise to His name. We shall ever feel the freshness of the morning and shall ever be far from its close. 'And they need no candle, neither light of the sun; for the Lord God giveth them light.' Revelation 22:5. The light of the sun will be superseded by a radiance which is not painfully dazzling, yet which immeasurably surpasses the brightness of our noontide. The glory of God and the Lamb floods the Holy City with unfading light. The redeemed walk in the sunless glory of perpetual day."[4]

The environment, system, and order experienced by the redeemed, though vastly different from what we experience under the second law of thermodynamics, is perfectly safe to the righteous—desirable, and preferable to our order. This is another evidence of the safety of reversing the order to disorder arrow in the second law of thermodynamics. But while the righteous are safe, what about the wicked? Would they all be killed if we should build a second law of thermodynamics reversing machine and flipped the switch? It is not our purpose to do that, but remove from the wicked their propensities to evil, and elicit from them new behavior in harmony with the law of God.

According to the Scenario A prophecies (the worst case scenario [see *Does God Really Love Us*, Chapter 10]), "at the coming of Christ the wicked are blotted from the face of the whole earth—consumed with the spirit of His mouth and destroyed by the brightness of His glory. Christ takes His people to the City of God, and the earth is emptied of its inhabitants."[5] The righteous are safe, but the wicked are destroyed. The glory of Christ is related in some way to the reversed second law of thermodynamics. If we reversed the order to disorder arrow in the second law of thermodynamics, would all the wicked be slain?

Before we answer that question, let us notice what happens a thousand years later to the wicked under Scenario A:

At the close of the thousand years, Christ again returns to the earth. He is accompanied by the host of the redeemed and attended by a retinue of angels. As He descends in terrific majesty He bids the wicked dead arise to receive their doom. They come forth, a mighty host, numberless as the sands of the sea. What a contrast to those who were raised at the first resurrection! The righteous were clothed with immortal youth and beauty. The wicked bear the traces of disease and death.[6]

The resurrection of the righteous and the resurrection of the wicked both have to do with reversing the order to disorder arrow in the second law of thermodynamics. But why such a difference? Why should the resurrected righteous be immortal, young, and beautiful, but the resurrected wicked be old, have traces of disease and death?

There are at least two ways to reverse the order to disorder arrow in the second law of thermodynamics. And they both are demonstrated here. One way is the way we have planned to do it—by fusing the anti-semions in positrons, and converting them to protons or neutrons. This method is relatively slow and gentle. The wicked could be resurrected by a process similar to backing up of the clocks. But if that process is terminated before the moral and physical restoration is complete, the wicked would be like just before they died, not immortal, beautiful youth.

The other process envisioned in the model to reverse the order to disorder arrow of the second law of thermodynamics is to fuse the constituents of octon-anti-octon pairs. Octon-anti-octon pairs are very similar to light, but have much more tendency to ionize the vacuum and fuse. These may be responsible for reversing the order to disorder arrow, but they are fast and hard. While they are life to the righteous, they brook no evil. They are death to the wicked. Their work is more instantaneous. But only God has control of octon pairs. It is not yet a thing available to the control of His creatures. But see *The Physics of Genesis* Draft 3 Chapter 3.)

Let us see the manner of the death of the wicked after their judgment. "By a life of rebellion, Satan and all who unite with him place themselves so out of harmony with God that His very presence is to them a consuming fire. The glory of Him who is love will destroy them."[7] Apparently the glory in the face of God is octal light—or octal fire. It is life to the righteous, but death to the wicked. But we would not reverse the order to disorder arrow in the second law of thermodynamics with octal fire. We would use the slower, safer anti-semion fusion (the constituents of positrons). We have considered nothing, so far, which would indicate that it would not be safe to operate our machine.

The Bible predicts a lot of miracles in the last days that can be attributed to reversing the order to disorder arrow in the second law of thermodynamics (Isaiah 11:6-9; Hosea 2:18; Isaiah 29:18; 35:6; 33:24; 58:12; etc.). Let us just focus now on the large ones that may be dangerous.

But tidings out of the east and out of the north shall trouble him: therefore he shall go forth with great fury to destroy, and utterly to make away many.

And he shall plant the tabernacles of his palace between the seas in the glorious holy mountain; yet he shall come to his end, and none shall help him.

And at that time shall Michael stand up, the great prince which standeth for the children of thy people: and there shall be a time of trouble, such as never was since there was a nation even to that same time: and at that time thy people shall be delivered, every one that shall be found written in the book.

And many of them that sleep in the dust of the earth shall awake, some to everlasting life, and some to shame and everlasting contempt.

And they that be wise shall shine as the brightness of the firmament; and they that turn many to righteousness as the stars for ever and ever. Daniel 11:44-12:3.

Does Jesus do these things? Or do we do these things at this time and under these circumstances by reversing the order to disorder arrow in the second law of thermodynamics?

Making human faces shine, like they were on fire, is an occasion of amazement and panic among the uninformed and wicked, according to Bible prophecy. Isaiah 13:8. However, it is harmless.

"Moreover the light of the moon shall be as the light of the sun, and the light of the sun shall be sevenfold, as the light of seven days, in the day that the LORD bindeth up the breach of his people, and healeth the stroke of their wound." Isaiah 30:26. The author first thought this was a one day transient during a nova of the sun—that it was a great danger to the human race. But Ellen White quotes this text and applies it to the steady state of the moon and sun for the redeemed.[8] That may be the effect of reversing the order to disorder arrow of the second law of thermodynamics for the sun, which our machine would be able to do. Before we do that, however, we may have a lot of test cases with smaller radii reversed. We can experiment with reversed radii as small as ten meters.

Reversing the order to disorder arrow in the second law of thermodynamics is not the same as reversing the clocks. The motions of operating engines, in some respects, are like the movements of a clock. When the order to disorder arrow is reversed, engines would not operate backwards. They would be un-aged and made safer. The reversal of the order to disorder arrow in the second law of thermodynamics would do little to our current modes of transportation except to make them safer. However, look at it this way: The glory that there would be from human bodies in that event may not be just light. It may be omni-directional coherent light suitable for inertia-less travel. The human form may become an inertia-less craft to replace cars, planes, and rockets. We may not need so much power generated.

In this chapter we have considered the safety of reversing the order to disorder arrow in the second law of thermodynamics. We have seen it would not only be safe, but far preferable for the righteous. For the wicked we have found it is safe provided it is by the fusion of anti-semions in positrons, which is the case with our proposed machine, not by the fusion of octon-anti-octon pairs. Nothing we have studied would discourage us from experimenting with this phenomenon in a small local area.

It would also be safe to reverse the order to disorder arrow on the sun. On the least part, it would back up the sun to an earlier safer time. At the greatest, it would make the sun brighter, but not painfully dazzling. It would be safe for previously wicked persons.

---

[1]Ellen G. White, *Patriarchs and Prophets*, p. 57.

[2]*Ibid.*, pp. 57-59.

[3]Genesis 3:7-19.

[4]Ellen G. White, *The Great Controversy*, pp. 675, 676.

[5]*Ibid.*, p. 657.

[6]*Ibid.*, p. 662.

[7]Ellen G. White, *The Desire of Ages*, p. 764; Deuteronomy 4:24; Hebrews 12:29.

[8]Ellen G. White, *My Life Today*, p.348.

Chapter 5

What Would the Refresher Do?

Reverse aging for adults

The simplest effect of the Refresher to understand is reversing adult aging. Old people can be made young adults again in the active footprint of the Refresher. This effect for positron anti-semion fusion does not really back up time or the clock. It merely reverses the order to disorder arrow in adults. It saturates at the maximum state of order—which is young adulthood. It reverses adult aging at a rate of about 1836 times as fast as the rate the original adult aging occurred. A century of aging can be reversed in just under 20 days (19.89 days).

Resurrections from the dead

The reverse aging occurs also for bodily remains—re-assembling dust and bones into living beings again. All the dead of all ages of earth's history would be resurrected in about 3½ years of Refresher machine time, starting with those who died most recently.

Backing diseases out of existence

In the process of reverse aging, diseases would be backed out of existence. This would work also for difficult diseases like HIV AIDS, cancer, cystic fibrosis, and Ebola.

Reversing all decay

Spoiled fruit would un-spoil in the active footprint of the Refresher. Fresh fruit would stay at the maximum state of order for fruit forever—fresh picked fruit. And this would be without refrigeration. This would amount to a new kind of food preservation without canning or freezing.

This process would un-decay everything in the Refresher footprint, not just fruit. And the footprint could be enlarged to cover the entire earth and sun.

Reversing pollution out of existence

In the Refresher footprint, all pollution would be reversed out of existence. Depending on the Refresher control settings, this effect could be world-wide.

"Raising up the foundations of many generations" Isaiah 58:12.

The Refresher would automatically rebuild previous decayed structures. It would rebuild and restore the entire earth.

Reversing forest fires

The Refresher not only would stop forest fires in its footprint, but would reverse the fires—restoring all that was lost—animate and inanimate, including lost trees and homes.

Reversing all calamities
Reversing all effects of war
Preventing all munitions from firing
"Making wars to cease to the end of the earth." Psalm 46:9.
Removing sinful propensities from people, including criminals
Emptying prison houses
Making possible and efficient Clean Energy Sources.
Taming the Sun

We could tame the sun, back it up to safer times, or operate it in the second law of thermodynamics reversed phase simply by putting the sun in the operating footprint of the Refresher, by having one positron collision and fusion every 79 hours. This can be done safely on earth. All the earth would be in the active footprint of the refresher when the sun is being rejuvenated.

The blessings of the Refresher are endless.

In short it would restore earth to Edenic perfection in about 3½ years of machine time.

Chapter 6

SHOULD WE DO IT?

Should we build and operate an order to disorder arrow in the second law of thermodynamics reverser (Refresher)? Apparently to do so would be safe. But why should we do it?

People are spending thousands upon thousands of dollars for medical diagnostics and treatments that do not cure, where they could have a complete cure for free if the Refresher were operating. More than that, they could have the resurrection of their dead loved ones for free if the Refresher were operating. The suffering and dying of people weigh upon the author—especially the suffering and dying of people interested in the building and operating of the Refresher.

But people have suffered and died for thousands of years. Why should that change now? Why not thousands of years ago? It is only recently that science has been advanced enough to build and operate the Refresher. And besides, science has taken some wrong turns, and is entrenched in wrong theories. This breakthrough would not be possible without the Spirit of the Everlasting Father indwelling a theoretical physicist and guiding the research. But why did not the Spirit of the Father incarnate Sir Isaac Newton or an earlier scientist? The Father could not personally incarnate a scientist until His glorious Person was consumed by man, and that could not be done until the invention of the atomic bomb and the dropping of them in inhabited cities. (Zephaniah 1:7, 8; 2 Peter 3:10-12.) The incarnated child (Isaiah 9:5-8) could not be born before 1946.

Why should this generation be the favored generation? Is that not unfair to previous generations? The human race is in a grand relay race. When last day scientists win, all previous generations win also, because the coverage of the refresher would not only expand to cover all the earth quickly, the dead from generation after generation in the past would be resurrected and incorporated into society. They would receive the blessing. In time, every last spec of life that was conceived or created would be restored. It would be a victory for the whole human and angelic races.

"You speak as though all the dead would be saved. Prophecy indicates 'a mighty host, numberless as the sands of the sea' would be lost."[1]

Jonah was instructed by God to preach, "Yet forty days, and Nineveh shall be overthrown." Jonah 3:4. Did it happen? No. Was the prophecy false? No. It was a conditional prophecy. It really would have been fulfilled if Nineveh had failed to repent. But all Nineveh, including the king, repented in sackcloth, ashes, and fasting. God spared the city.

Jonah lost sight of the interests of others, and was more concerned with being regarded as a false prophet. Are not Seventh-day Adventists in danger of the same thing? For over one hundred years they have preached *The Great Controversy* scenario of Bible prophecy. If billions are not lost, Seventh-day Adventists are in danger of concern that they are false prophets—that Ellen G. White was a false prophet—that God lied. But Ellen White plainly said, "It should be remembered that the promises and threatenings of God are alike conditional."[2] The Bible and Spirit of Prophecy predictions of the end of the world, the millennium of Revelation 20, and hell fire were conditional on there being only one Sacrifice for sins. God offered it to them that way on at least two occasions (1844[3] and 1888) but God's people failed Him. God was forced to make the second Sacrifice for sins in 1945. This changes everything! As Jesus by His Sacrifice bought the sinner another trial,[4] just so the Father purchased for the wicked lost—those that squandered the trial purchased for them by Jesus—another trial—a third trial. They need not be eternally lost—not even sinners long dead, like the antediluvians. They are all resurrected to have another trial.

However, no living or resurrected dead sinners shall receive immortality before they come into harmony with the law of God. With positron fusion, they are resurrected with traces of disease and death, not the immortal beauty and youth of the righteous. However, the longer they stay in the field, the more healed they will be, and the closer to the maximum state of order they will be in young adulthood. If they stay in the field, eventually they will have no more propensities to sin. Jeremiah 31:31-33; Ezekiel 36:25-27. Theoretically a person could choose still to sin even then and thus stay mortal. But why? That person's every inclination would be to do right. Would not everyone in time have the victory over sin? Would not every speck of life be saved? That is, if we build and operate the machine. Should we not do it?

For about two hundred years, Bible believing Christians have been preaching that Jesus is coming soon. They have endeavored to get ready for that event. But Scripture says, "And he shall send Jesus Christ, which before was preached unto you: Whom the heaven must receive until the times of restitution of all things, which God hath spoken by the mouth of all his holy prophets since the world began." Acts 3:20, 21. Maybe the coming of Christ is on hold until the restitution of all things through the operation of the refresher machine. Should we not build and operate it?

Like Jesus before him, the anointed one in this generation "shall cause the sacrifice and the oblation to cease." Daniel 9:27. Jesus caused the animal sacrifices to cease, and replaced them with His high priestly ministry based on His Calvary Sacrifice. But when the coverage of the Refresher field is worldwide, and the Father's Sacrifice is effective for all men, Jesus need no longer offer His blood in the Most Holy Place in heaven, but He can come to earth to receive the throne of His father David.[4] Shall we not build and operate the Refresher machine?

Jesus has raised His hand to heaven and sworn by Him that liveth for ever and ever "that there should be time no longer: But in the days of the voice of the seventh angel, when he shall begin to sound, the mystery of God should be finished, as he hath declared to his servants the prophets." Revelation 10:6, 7. Prophetic time ceased in 1844 at the end of the longest time prophecies. Literal time would cease in the field covered by the operation of the Refresher machine. Jesus swore it would. Shouldn't we do it?

40

"And the seventh angel sounded; and there were great voices in heaven, saying, The kingdoms of this world are become the kingdoms of our Lord, and of his Christ; and he shall reign for ever and ever. And the four and twenty elders, which sat before God on their seats, fell upon their faces, and worshipped God, saying, We give thee thanks, O Lord God Almighty, which art, and wast, and art to come; because thou hast taken to thee thy great power, and hast reigned." Revelation 11:15-17. This is the seventh angel referred to in Revelation 10:7. All heaven, including the twenty four elders, are ready to celebrate when our Lord receives the kingdoms from this world only to give them in turn to his Christ—or Jesus Christ. The Lord in this day is God the Father incarnate, and the great power that he takes unto himself is the positive order energy released from the order to disorder arrow reverser (Refresher). The physicians in Jesus' day received large payments for treatments that did not cure. When Jesus healed everyone for free, the people wanted to make Him King. The same will be true in the near future. Multitudes upon multitudes will be healed for free. They will be reunited with loved ones lost through death. They will be delivered of all pain and sorrow. Therefore they will want to make God's man King. He will acquiesce only to turn the government over to Jesus.

God's man has given this government a lot of thought. He has concluded that deep in the heart of every person—including the worst criminal—is a beautiful, valuable person. But everyone has been beat up by the second law of thermodynamics—in different ways. Some have lost their health. Some have sinned greatly. But all have been beat up by the second law of thermodynamics. The reward of the righteous should be to be liberated from the bondage of the second law of thermodynamics. The sentence pronounced against the wicked should be just the same—free them from the cruel bondage of the second law of thermodynamics, and release their beautiful person inside. With this philosophy of administration, God's man has the prophesied "sure mercies of David." Isaiah 55:3.

Without the creation and operation of a Refresher quickly, the sun will peak seven times brighter and hotter soon, killing

41

millions of people. Shouldn't we fund and build a Refresher quickly? Even if we survived the nova peak without a Refresher, the sun would soon go out in darkness. This would be far worse than the nova peak! If we wanted to survive this, we should not only make one or more Refreshers, but mass produce Clean Energy Sources on earth, making the light and heat of the sun unnecessary. But what we do, we need to do quickly!

[1]Ellen G. White, *The Great Controversy*, p. 662.

[2]Ellen G. White, *Selected Messages*, Book One, p. 67.

[3]Ellen G. White, *The Great Controversy*, pp. 457-458.

[4]*Ibid.*, p. 416.

Chapter 7

The Messiah's Inventions

Noah's ark was an invention of God (elohiym—the plural (three) form of one God—God the Father, God the Son, and God the Holy Spirit). God the Son later became the incarnate Jesus of Nazareth, the Messiah, the Anointed One, the very Christ. The original Son of God was undoubtedly the divine agent that communed with Noah, inspiring him with the great invention of God to preserve life through the perils of a world-wide Deluge. Noah was the master builder and preacher of righteousness.

God could have translated the people and animals to heaven or some other world before that horrific water holocaust of the entire earth, and relocated them back to earth when the Flood was over. But God chose not to do this. God chose a cooperative effort of God and man at great cost to preserve life on earth during that fearful water holocaust. But the ark, as well designed and built as it was, was in itself not sufficient to preserve life. It required the mighty power of God and the heavenly angels to guide and preserve the ark and its inhabitants during that fearful ordeal.

"For this they [last day scoffers—evolutionists] willingly are ignorant of, that by the word of God the heavens were of old, and the earth standing out of the water and in the water: Whereby the world that then was, being overflowed with water, perished: But the heavens and the earth, which are now, by the same word are kept in store, reserved unto fire against the day of judgment and perdition of ungodly men." 2 Peter 3:5-7 and context (KJV).

In the imminent future we face the fearful peril of one third of the population of the earth being incinerated in a nuclear holocaust (Revelation 9:14-21); also the peril of the corona of the sun going out in darkness, like what happened briefly July 19-23, 2013 (Google dark sun), and the earth's population freezing to death, or the sun continuing its nova sequence, scorching the earth with great heat, and then going out in darkness. What we desperately need now are new inventions of God, the cooperation

43

of God, men, and angels to preserve life in fire, make the sun unnecessary, flood the entire earth with light and clean energy (Revelation 21:23).

God has already inspired one man, said of Lucifer himself to be the Everlasting Father incarnate, the Messiah, the Anointed One, the very Christ, with designs of God of inventions to preserve and restore earth to Edenic condition in this doom of fire. The inventions are already roughly designed and the master builder found and educated. All that is needed now is about $10,000,000 more and the chance to go right to work building these heavenly inventions. God will not do this all by Himself. He will work with us as we cooperate with Him.

Chapter 8

Correcting the Standard Model of Physics

A. Impossible with Einstein's Special Theory of Relativity

The Theory of the Refresher is impossible to derive with Einstein's Special Theory of Relativity. Also it is impossible to get a government or private grant to build a Refresher unless the reviewers are thoroughly un-deceived from Einstein's Special Theory of Relativity. And that is impossible to do in a 25 page grant application. It is necessary to take time out to make a grand synthesis of physics and spiritual truths like this tome to break down the barricades to scientific progress by science falsely so called.

There are several faults with Einstein's Special Theory of Relativity: He tried to do everything without an aether. It is impossible to have a Universe that does not either blow up or collapse on itself without an aether. But with an aether, it is simple—just spin the whole Universe. But that is impossible to define without an aether. Scientists still do not have an acceptable model of gravity and inertia. It was easy for me 31 years ago with an aether (see *Electrino Physics,* Chapter 5). That work even united gravity and inertia in one formula—the beginning of a Unified Field Theory, which Einstein tried for 30 years to achieve and could not achieve, but which I achieved at least seven years ago. (See *Electrino Physics*, Chapter 7, "Uniting the Forces".)

How did I know there was an aether so many years ago? I exercised my God given special gift of abstract reasoning and pattern recognition on subtle differences between different models. But most of all, my best evidences were from the Bible. At Upper Columbia Academy, in the Spring of 1964, the subject of an aether came up in the physics class. Was there, or was there not an aether? Einstein said no. Some other scientists said yes. Who was right? The students in our class were polled, "is there, or is there not an aether?" Before we made up our minds, our teacher Robert R. Ludeman shared with us the best evidences consisting of a list of experiments bearing on the subject and their purported results. The night following was the full moon in April, 1964. As I lay in bed, I mentally modeled the particle reactions in the listed experiments. My verdict: "there is an aether, Einstein notwithstanding." In after years I never varied from that conclusion.

My positive stand, however, was noised among the class's students, some of whom heckled me. "Do you think you are smarter than Einstein?" "How do you know it is possible to go faster than the speed of light?"

"Because the angel Gabriel flew from heaven to Daniel in the time it took Daniel to pray the prayer recorded in Daniel 9. When I read the prayer slowly, it took me less than five minutes. From heaven to earth in five minutes! How far away do you think is heaven? Less than one light year? I don't think so. Doesn't Ellen White (messenger to the remnant) indicate heaven is in Orion (1,350 light years away)? But let's just say heaven is one light year away. Traveling one light year in five minutes! That is much faster than the speed of light!"

But some of the hecklers still reasoned, "But Gabriel is an angel, and angels are spirits. Maybe spirits can go faster than the speed of light, but flesh and bone people cannot."

But I said, "Then this case is settled by the experience of Jesus on the day of His resurrection. Shortly after His resurrection He ascended to consult with His Father. He then descended to meet with His disciples on the Road to Emmaus and in the Upper Room that evening. The disciples thought that He was a spirit. But Jesus said, 'Spirit hath not flesh and bone as you see me have.' Even in His resurrected state Jesus had flesh and bone, which broke the speed of light barrier on the trip to and from heaven."[1]

In his Special Theory of Relativity, Einstein said it was impossible for anything to go faster than the speed of light. But recently neutrinos were clocked going faster than the speed of light.[2-4] Some have disputed those results, attributing them to various experimental errors. But it is impossible to unite all the forces in a Unified Field Theory without some particles traveling faster than the speed of light.[5]

There are several other difficulties with other aspects of Einstein's Special Theory of Relativity such as time slippage, reversing the real and the apparent, and angle of measurement.

Special Relativity's Weaknesses

Except for popularity and preeminence in scientific publications, Einstein's aetherless theory of relativity has not done quite as well as Lorentz's aether model of relativity. Without any preferred reference frame, Einstein's theory has not accounted for the experimental data as simply as Lorentz's theory with a preferred reference frame. Scientists can argue that Einstein's theory can account for all the data. But other scientists can also argue that it cannot account for all the data parsimoniously. The parsimony principle is an accepted principle of physics. It is a postulate in *Electrino Physics* in Chapter 6. This principle should have weight in this contest.

Another area of difference between SR and LR is the twin or clock paradox. In the twin paradox, one of two twins rockets away from earth a distance into space, turns around, then rockets back. To the traveling twin, the stay at home twin appears to make a round trip in the opposite direction. Which one will age more than the other, or will they age alike? Will each twin claim that the other aged less because of time dilation? Or will there be a difference in the aging? In SR there is a twin or clock paradox.

How does the resolution of the "twins paradox" compare in LR and SR?

In LR, the answer is simple: The Earth frame at the outset, and the dominant local gravity field in general, constitutes a preferred frame. So the high-speed traveler always comes back younger, and there is no true reciprocity of perspective for his or other frames.

In SR, the answer is not so simple; yet an explanation exists. The reciprocity of frames required by SR when Einstein assumed that all inertial frames were equivalent introduces a second affect on "time" in nature that is not reflected in clock rates alone. We might call this effect "time slippage" so we can discuss

47

it. Time slippage represents the difference in time for any remote event as judged by observers (even momentarily coincident ones) in different inertial frames.[6]

The author of the above quote goes on to give several numerical examples of time slippage for the traveling twin when he turns around, in order that complete reciprocity of frames may be maintained in Einstein's Special Theory of Relativity. Lorentzian Relativity doesn't need any time slippage. It has a preferred frame. Perhaps scientists should make one more unmanned lunar orbit mission with atomic clocks and return of the module to earth for comparison of the atomic clocks with ground clocks, to see if there is any time slippage, or to see if the clocks are as in LR with a preferred reference frame.

The problem is caused in Einstein's SR because he did not believe there was any way of determining an absolute space in special relativity. He postulated that one uniformly moving reference frame was as good as another. The theory of relativity was thought to be consistent no matter what the rest velocity was assumed to be. Therefore twin A should experience length contraction and time dilation as computed by twin B if twin B assumes he is at rest during his flight.

Some physicists try to harmonize the clock paradox through general relativity, as though the accelerations of twin B affect the positions of his clock. "Cyclotron experiments have shown that, even at accelerations of 10^{19} g (g = acceleration of gravity at the Earth's surface), clock rates are unaffected. Only speed affects clock rates, but not acceleration per se."[7]

The clock paradox can be put in better perspective by studying triplets instead of twins. Let triplet X stay at home on planet earth. Let triplet Y rocket out in space similar to twin B, and let triplet Z rocket out in space in the opposite direction as triplet Y. After a period of coasting, let triplets Y and Z simultaneously decelerate and rocket back to earth for a close-encounter fly by the earth and each other. Let X, Y, and Z compare their clocks. Then Y and Z should continue to coast for awhile, then simultaneously decelerate and rocket back to earth for another close-encounter and fly by. Let them again compare their clocks. Finally let Y and Z continue coasting for awhile, simultaneously decelerating and accelerating back to planet earth, where they land and compare their clocks with triplet X.

Who should have slower clocks? X, Y, or Z? Z should expect Y to have time dilation relative to him according to Einstein's theory. Also Y should expect Z to have time dilation relative to him. One cannot appeal to the accelerations and general relativity to harmonize this contradiction, for the accelerations are symmetrical. Triplet X would expect Y and Z to be time dilated equally.

None of this is a problem in LR, which has a preferred reference frame.

Two Theories of Relativity

In 1904, one year before Einstein published his Special Theory of Relativity, Hendrik Antoon Lorentz originated a theory of relativity.

We must make a distinction between Einstein SR and Lorentzian Relativity (LR). Both Lorentz in 1904 and Einstein in 1905 chose to adopt the principle of relativity discussed by Poincare in 1899, which apparently originated some years earlier in the 19th century. Lorentz also popularized the famous transformations that bear his name, later used by Einstein. However, Lorentz's relativity theory assumed an aether, a preferred frame, and a universal time. Einstein did away with the need for these. But it is important to realize that none of the 11 independent experiments said to confirm the validity of SR experimentally distinguish it from LR—at least not in Einstein's favor. However, the issue of the need for a preferred frame in nature is, charitably, not yet settled. Certainly, experts do not yet agree on its resolution. But of those who have compared both LR and SR to the experiments, most seem convinced that LR more easily explains the behavior of nature.[8]

In Einstein's Special Theory of Relativity, relative mass increase, time dilation, and length contraction are only *apparent*—observational phenomena due to rotation in complex space-time. In quasi-relativity, mass increase, time dilation, and length contraction are *real*—caused by motion relative to an aether. The resultant constancy of the speed of light in quasi-relativity is only *apparent*, not *real* as in Einstein's special relativity. Quasi-relativity reverses what is considered *real* or *actual* lengths and times and what are *apparent* lengths and times. The mass, clock speed, and length of a moving observer as seen by himself only appear to be the rest mass, rest clock speed, and rest length. They are not actually rest quantities, and invariant, as in Einstein's theory.

Transformations in quasi-relativity are different than in special relativity. To transform position, time, momentum, or energy (mass) to another frame in quasi-relativity, one must first make an inverse transformation to the aether rest frame according to the direction and magnitude of the observer's velocity relative to the aether, then make an appropriate rotation of the x axis to the direction of the velocity of the target frame relative to the aether, then make a transformation according to the magnitude of the velocity of the target relative to the aether. Every correct transformation should be a triple transformation like this. Such a triple transformation [1] an inverse transformation of increased clock speed, decreased mass, and length expansion when going from the observer to the aether rest frame; 2) an appropriate rotation; and 3) a transformation of decreased clock speed, increased mass, and length contraction when going from the aether rest frame to the other arbitrary frame, where all transformations go by the clock in the aether frame] is not in general equal to either a single transformation or a single inverse transformation. Here is where Einstein gets into trouble with the clock paradox. Not every uniformly moving frame is equivalent. Relative velocities between observer and target can vary from 0 to ± 2V, with time and energy (mass) remaining constant.

The author derived *Relativity in an Ether*[aether] in 1977 when he did not know of Lorentz' work or anyone else's theory of relativity except Einstein's. Today derivations of relativity are a dime a dozen. Thousands of scientists dissent from Einstein's theory of relativity. But their derivations require ad hoc discontinuous hypotheses to reverse the natural relativistic effects in the Galilean transformations. The author's derivations are still the best, employing Newton's third law as a postulate for a smooth continuous derivation from the Galilean transformations. See *Electrino Physics*, Chapters 1-8 and Problem Solutions, Chapters 1-8 (relativity derivations, start of a theory of gravity and inertia, Unified Field Theory, beginning of a Unified Particle model, and Unified Universe derivations), http://benevolententerprises.org Book List. Please observe that the first chapter noted may seem to be boring irrelevant scientific lore. But it is necessary to establish a natural discovery pattern for our abstract reasoning and pattern recognition to tackle the relativity mystery. All files listed may be downloaded free at http://benevolententerprises.org Book List. More up-to-date *Electrino Physics* Draft 2 may be obtained at createspace.com, amazon.com, barnesandnoble.com, and 2,500 book stores.

Another difference of the GUT to the Standard Model is that charged sub-particles of like fusion states can fuse to particles of higher fusion states.[9] The secret of why that should be is that when sub-particles orbit or travel faster than the speed of light in the relativistic frame relative to the baseline non-relativistic frame, their radii become imaginary because of the relativistic length contraction formula. The strong electric force equation for these super luminal sub-particles has two such imaginary radii multiplied together. That makes an additional minus sign in the force equation, which makes like charges attract. When two bound sub-particles of a positron collide with 1880 MeV energy or more with two other bound sub-particles of another positron with like oriented spins in the Center of Mass Frame, the four sub-particles are attracted into the same orbit. Then one sub-particle from one positron is more attracted to one sub-particle from the other positron than to any other sub-particle because of closer proximity. These are like charges, and here like charges attract because they travel faster than light. The two sub-particles are attracted by the electric strong force. Nothing stops them from fusing. The other two positron sub-particles fuse also. Four sub-particles fuse down to two particles each with twice the charge as the charge of one of the four sub-particles. The four sub-particles are ½ e charges. The two fused particles are 1 e charges each, but though they are numerically whole particles, they cannot exist alone. That is why on creation they scavenge from the graviton sea the necessary sub-particles to become protons or neutrons. But when the positron four ½ e sub-particles fuse to the two 1 e particles, they switch from antimatter to matter. The fusion of sub-particles in positrons results in the generation of solely positive order energy (quantum mechanical energy in the creation of particles). This phenomenon is theorized to reverse the order to disorder arrow in the second law of thermodynamics [because it is positive order energy as opposed to negative order energy which surrounds us and which determines the current order to disorder arrow direction and the direction of reactions].[10]

[1]Gordon L. Ziegler, *Seeking a Unified Universe*, Chapter 4, "Upper Columbia Academy," (http://benevolententerprises.org Book List.)

[2]Frank Jordans and Seth Borenstein, "Neutrinos clocked moving at faster-than-light speed" (Associated Press): http://www.msnbc.msn.com/id/4462927/ns/technology_and_science-science/?gt1=43001.

[3]Robert Evans, "Particles found to break speed of light" (Reuters): http://www.reuters.com/article/2011/09/22/us-science-light-idUSTRE78L4FH20110922.

[4]"Do neutrinos move faster than the speed of light?"— physicsworld.com: http://physicsworld.com/cws/article/news/47283.

[5]Gordon L. Ziegler, *Electrino Physics*, Chapter 7 (http://benevolententerprises.org Book List.)

[6]Tom Van Flandern, Univ. of Maryland and Meta Research, Open Questions in Relativistic Physics, edited by Franco

Selleri (Montreal: Apeiron, 1998), pp. 81-90, as quoted by "What the Global Positioning System Tells Us About Relativity, Meta Research, http://www.metaresearch.org/cosmology/gps-relativity.asp.

[7]C. MØLLER, *The Theory of Relativity* (Oxford: Clarendon Press, date unknown to author).

[8]Tom Van Flandern, *op. cit.*

[9]Gordon L. Ziegler, *Electrino Physics*, Chapter 12.

[10]*Ibid.*, Chapter 16.

B.   Impossible With Electrons as Spinning Point Charges

Post-Modern Electrons

By Gordon L. Ziegler

Abstract

The modern electron theory was put forth by P. A. M. Dirac in 1928—namely that electrons were spinning point charges. Dirac was satisfied with only a first order approximation fit with measurements—which has been adequate for all the modern era. But very high energy electron collisions in the post-modern era require more precision than that. This paper will identify what was wrong with the spinning point charge theory and how to correct it.

Electrons cannot be spinning point charges in the absolute, because a point charge in the absolute has an infinite mass, and an electron has a small mass. Furthermore, the relativistic radius of the electron is not real but imaginary. The concept of particle spin models after $mrc = n\hbar$. We can measure the particle mass m and the radius r, but how do we measure the rate of the rotation of the surface of the charge? We could shine a light of some color on the edge of the spinning charge to see what bounces back to determine the speed of rotation. But notice this works only if the charge surface is bumpy or lumpy. If the charge surface is smooth symmetric, the light will just go by without reflecting. This is at the elemental level. Therefore we can take from this a far reaching postulate—a smooth symmetric charge distribution cannot have detectable spin. It can have spin, but not detectable spin. Common reactions in particle physics model after detectable spin.

A spinning point charge is smooth symmetric. Therefore, if there would be such a thing, it should have no detectable spin. But electrons have detectable spin. Therefore electrons should not be spinning point charges. They should be bumpy or lumpy. The author's model for the structure of the electrons is two half charges orbiting each other at the speed of light. Such a system is lumpy, and it spins. The orbit of the half charges at the speed of light means the half charges are bound in a speed of light barrier. They cannot be blasted apart. They act as one particle. This fact negates the common objection that a two particle electron should be capable of being blasted apart, but electrons always act as single particles.

Various references on spinning point charge electrons and P. A. M. Dirac authorship of the idea are in [1-4]. That Dirac was satisfied with only a first order fit to the measurements is referenced in [4].

---

[1] geocalc.clas.asu.edu/html/GAinQM.html.

[2] en.wikipedia.org/wiwi/Spin_(physics).

[3] en.wikipedia.org/wiki/Electron_optics.

[4] The Quantum Theory of the Electron, P. A. M. Dirac, Proceedings of the Royal Society of London. Series A, Containing Papers of a Mathematical and Physical Character is currently published by The Royal Society.

C.   Impossible With Murray Gel Mann's Quark Hypothesis

1.   How many is too many?

There is a law in science (parsimony principle) that things should be made of as few different kinds of elementary particles as possible. Old science (the Standard Model of Physics with the Quark Hypothesis) requires 61 different kinds of elementary particles to put together light, matter, but **not** gravitons. New science (the Electrino Fusion Model of Physics with the Electrino Hypothesis) requires only one kind of elementary particle to make light, matter, **and** gravitons. Which do you think has the better science—old science or new science?

2. Matching spins and charges

Old science goes by the Quark Hypothesis, which has ⅔ charge and ⅓ charge for the smallest charges to make up everything. They try to match these irregular charges with ½ spins. It is not a very good match. Maybe that is why it takes old science 61 particles to make up light and matter.

New science goes by the Electrino Hypothesis, which has 0 spins for each of the electrinos (1 charge, ½ charge, ¼ charge, and ⅛ charge) and ½ spin for their minimum detectible orbital spins. Doesn't new science have a better match of spins and charges than old science? Maybe that is one reason why new science can make everything out of only one particle.

D.   Explaining Things No Other Theory Can

Electrino Fusion Model versus the Standard Model
Parsimony Principle: It takes the Standard Model of Physics 61 different elementary particles to compose light, matter, but **not** gravitons. It takes the Electrino Fusion Model of Physics only one particle to compose light, matter, **and** gravitons.
Uniqueness: The Standard Model has seven structure formulae for eight particles—the first and the last are the same.

But the masses are not the same! The Standard Model does not have a unique structure formula for each particle. The Electrino Fusion Model does.

You cannot make gravitons out of quarks, but you can make them out of electrinos. The Electrino Fusion Model solves for four different gravitons on the lowest chonomic level.

The Standard Model has g/2 factors for two particles. The Electrino Fusion Model has g/2 factors for 28 particles derived from first principles.

The Standard Model cannot derive the masses of light particles from first principles. Those must be input in their model. The Electrino Fusion Model has derived the masses of 22 light and heavy particles to two to four place accuracy from first principles. We are working on the rest. We have a book published explaining the method of calculating the masses of the 22 particles from first principles—see *Advanced Electrino Physics* Draft 2 (orderable through amazon.com, barnsandnoble.com, or createspace.com). See also *Predicting the Masses*, Volume 1, Introducing Chonomics from the above on line book stores.

Chapter 9

Clean Energy Sources

Electrino fusion can not only supply perfect health, soundness of mind and body, eternal life, and the resurrection from the dead, it can supply the human race and the Universe with free, absolutely clean energy—no Carbon emissions, no radioactive wastes, no wastes at all, and little or no heat pollution! It is 1000 times as efficient as nuclear power; it can go 100 or 200 years without refueling; it doesn't require dangerous radioactive fuels or dangerous chain reactions. It can safely be shut off at a flip of a switch with no decay heat. It has no hazard of a meltdown or a radiation accident. It is very efficient. The annihilation of a single penny would produce $1,872,000 worth of electricity at bargain electricity rates of $0.03/kwh. The first model would use inexpensive common brass for fuel. Whereas a new nuclear power plant would cost $6.0 billion and have 1250 MW AC output, Electrino Fusion Power (EFP) Power would cost $110 million for the prototype and $50 million for successive plants and would produce 1880 MW each plant DC which could be converted to AC and put on existing electricity grids. Whereas a new nuclear power plant would take 10 years to build, the EFP power plant would take only one year to build or less and test, and could be pre-fabricated and proliferated around the world quickly.

Apart from the construction costs, EFP energy could operate for free with a few thank offerings. Were the construction financed by loans, the loans could be paid fourfold with $0.03/kwh in one year of operation, then free electricity afterwards.

To replace the solar energy on earth, to save the earth if the sun goes out at the end of a nova, we would have to produce 4000 times as much energy as the total earth electric production currently. And this would have to be done quickly. We would have to manufacture mass produce the Clean Energy Sources and proliferate them around the earth quickly, with training conventions of engineers, financiers, laymen and children in the great centers of population around the earth.

Chapter 10

Clean Energy Theory

1. Introduction

The technical name for the Radioactive Waste-free Reactor is the Electrino Fusion Power Reactor (EFP Reactor). Electrino is the author's name for tiny electric particles that compose all light, matter, and gravitons in the author's new Grand Unification Theory (GUT). The main difference between the Standard Model and the new GUT is that fracton charges in the GUT come in $\pm$ e, $\pm$ e/2, $\pm$ e/4, and $\pm$ e/8; whereas fracton charges in the Standard Model come in $\pm$ 2e/3 and $\pm$ e/3. The change in fracton charges did not lead to untenable particle structures. The author induced the structures of every known particle according to the scheme in the GUT. They all worked out all right. And whereas it takes 61 elementary particles to build known light and matter in the Standard Model, it takes only one according to the GUT. The GUT has deeper levels of symmetry and lower orbits. This chapter develops the features of the radioactive waste-free EFP Reactor using the new GUT.

2. Elementary Particle Fusion

In the new GUT (which, by the way, is called Electrino Fusion Model of Elementary Particles), the particles are held together by symmetrical orbits, not glued together by gluons. The quarks, with $\pm$ 2e/3 and $\pm$ e/3 fracton charges, do not lend themselves to stable, symmetrical orbits, but the electrinos, with $\pm$ e, $\pm$ e/2, $\pm$ e/4, and $\pm$ e/8 fracton charges, do. In the model, photons are composed of heavy positive and negative whole charges orbiting about each other, and traveling together at the speed of light; electrons are made up of like light half charges orbiting about each other; and pions are made up of two orbiting pairs of like light fourth charges orbiting in the opposite directions, superimposed on each other.

Notice the symmetry. Notice the orbits. Notice the space between the particles. Notice the individuality of the particles—bound only by the speed of light barrier and orbital mechanics.

It is important to notice the velocities of the particles and their behaviors at those velocities. All fractons (called electrinos in the model) travel either just slightly faster than the speed of light, or significantly faster than the speed of light. The point is, they all travel faster than the speed of light. For the light ones, this affects their radii—making them imaginary. This affects their force. Whereas slow like-charges repel, faster than c like-charges attract. This affects the potential energy of particles. This makes deep potential wells at the top of potential hills for the potential energy of charged particles. This affects the perceived mass-energy of the particles—positive instead of negative.

Faster than c like-charges attract. Negatively charged like half charges traveling just faster than c orbit around each other forming electrons. If the electrons never collide with any other electrons—at least not with sufficient energies—the half particle inertias in them cause the half charges to orbit always opposite each other—never approaching each other. But if electrons collide with each other with over 938 MeV each in the same orbit, four half charges come near to each other. The four half charges are not all held opposite each other. They all attract each other. What will happen? One half charge from one electron will be attracted to one half charge from the other electron. Nothing will stop the half charges. They will travel until they contact each other. What happens then? They are like charged. They form a new particle with twice the half charge—in other words a whole charge. We could say the half charges fuse to a whole charge.

When high energy electrons collide, not only do two half charges from opposite electrons fuse, the other two half charges on the opposite side fuse. We have four half charges from two electrons fusing to two whole charges. What then?

It is profitable at this juncture to assign fracton or electrino structures to simple particles. Pions are composed of four positive fourth charges—two orbiting one way, the other two orbiting the opposite way, superimposed on each other. Electrons are made of

59

two light weight negative half charges. Neutrons are constructed of a heavy positive whole particle orbited by an electron. If the constituents of pions were fused to the constituents of electrons, it would be to positive electrons—positrons—antimatter. If the sub-particles of negative electrons were fused to the heavy whole core particles of neutrons, it would be to negative neutrons—antimatter. If we started with the opposite charges of above, the particles would fuse to matter instead of antimatter. Every time there is a fusion of electrinos, there is a switch from matter to antimatter or vice versa.

What would happen to the negative half charges in electrons fused to whole particles above? The half charges would be negatively charged matter. The whole charges would be negatively charged core particles of antimatter—anti-protons and anti-neutrons. The anti-core-particles would scavenge from the graviton sea the remaining portions of anti-protons and anti-neutrons. The resultant anti-protons and anti-neutrons would drift into local protons and neutrons and annihilate them, giving off gamma rays, which could be converted into electricity. This is the foundation of the science of the radioactive waste-free EFP Reactor. The electricity comes from processed gamma rays, which come from the annihilation of protons and anti-protons and neutrons and anti-neutrons, which come from anti-protons and anti-neutrons, which come from negative heavy whole core particles (antimatter), which come from the fusion of half particles in electrons, which come from the collision electrons above 938 MeV each electron, with like spins in the center of mass frame.

3. Efficiencies

Before electrons can have fusion of their half particles, they must be accelerated to at least the masses of protons—938.27231 ± 0.00028 MeV [1]—roughly at least 939 MeV. That is a necessary energy investment into the process. When the particles fuse, there follows an annihilation of both a proton and an anti-proton or a neutron and an anti-neutron. Nearly twice as much energy in gamma rays results as was invested in the acceleration of electrons. At first this sounds good. But then we realize we must be more than 50 per cent efficient over-all in order to be self-sustaining and be an energy source using this energy phenomenon. That is hard to achieve. State of the art accelerator efficiency in 1988 was itself only 50% [2]. While individual steam turbine efficiencies were as high as 96.1%, the world record steam turbine gross efficiency recently was 48.5% [3]. That is an overall efficiency for our process of less than 24.25%. And we need 50% to break even, let alone have a surplus to become a new power source!

4. A Surprising Turn

The lack of necessary efficiency of the fusion-annihilation reaction is discouraging. The author put this process on the back burner until he would receive greater light upon the subject. Things took a surprising turn. Through fusing the sub-particles of positive electrons—positrons—in theory, he learned how to reverse the order to disorder arrow in the second law of thermodynamics. That is huge! That is a way to reverse aging, disease, and decay processes—to make old people young again and back out all diseases from existence! Let us read what he first wrote about the process and the phenomenon.

"The explanation that is usually given as to why we don't see broken cups gathering themselves together off the floor and jumping back onto the table is that it is forbidden by the second law of thermodynamics. This says that in any closed system disorder, or entropy, always increases with time. In other words, it is a form of Murphy's law: Things always tend to go wrong! An intact cup on the table is a state of high order, but a broken cup on the floor is a disordered state. One can go readily from the cup on the table in the past to the broken cup on the floor in the future, but not the other way round.

"The increase of disorder or entropy with time is one example of what is called an arrow of time, something

61

that distinguishes the past from the future, giving a direction to time." [4]

5. Electrino Model and 2nd Law

The natural tendency of leptons in beta decay is that the parent lepton combines with one or more gravitons to produce more particles. In all natural reactions, the order energy of the resultant particles is less than or equal to the order energy of the original particles.

a. Negative Energies. Let us consider antimatter more carefully. "In the Dirac theory also, *the permissible energy values for a free particle range from* $+mc^2$ *to* $+\infty$ *and from* $-mc^2$ *to* $-\infty$. The first of these results is of course just what we expect for a free particle—that its total energy can have any value greater than its rest energy. But the second result is quite puzzling, since it implies the existence of states of *negative total energy.*" [5] Anderson in 1932 discovered positrons in cosmic radiation. These were regarded as Dirac's negative energy particles. "The first two solutions of the Dirac equation . . . clearly describe a free electron of energy E and momentum **p**. The two negative energy electron solutions . . . are to be associated with the antiparticle, the positron." [6]

However, in the annihilation it is not $(+mc^2) + (-mc^2) = 0$, but $2mc^2$ is the result of annihilation. [7] There is something strange going on with the minus signs in these equations. The calculations are inconsistent.

Maybe there are two kinds of energy considered. One we can call entropy energy E_S. In the annihilation reaction, $|+mc^2| + |-mc^2| = 2mc^2$. Entropy energy is the higher value. The other energy is order energy E_O. In order energy the same reaction is $(+mc^2) + (-mc^2) = 0$.

Let us consider entropy energy and order energy for particle decay schemes. There are a few decay schemes where no negative order energy (anti-matter) is introduced in the right hand side of the decay schemes. In those few instances, the final order energy is equal to the initial order energy (when kinetic energy is taken into account). But in most cases, a trace of negative order energy (anti-matter) is introduced into the right side of the decay schemes. There is nothing on the left hand sides of the decay schemes to correspond to this addition of a trace of negative order energy on the right sides of the decay schemes. Therefore, total order energy is less on the right hand sides of the decay schemes than on the left hand sides (if only by a trace). A few decay schemes introduce a lot of antimatter (as K^-) on the right side of the decay scheme. The loss of order energy in the systems is greater in those cases. But in every case, for all natural processes, the order energy final is < the order energy initial, or

$$\Delta E_0 \leq 0. \qquad\qquad (10\text{-}1)$$

Let us check the order energy for electron electrino fusion reactions. Electrons made energetic by acceleration (as heavy as protons) fuse and form anti-protons. Matter is converted to anti-matter. Entropy energy is conserved, but not so order energy. Order energy is reduced in the extreme from +938 MeV to -938 MeV or more for each electron fused (two electrons are fused in each reaction). The order-disorder arrow for electron electrino fusion points in the usual direction. The system does obey the second law of thermodynamics as we now know it.

2.   Reversing the Order to Disorder Arrow. What would happen if we fused the electrino constituents of positrons instead of the electrino constituents of electrons? Entropy energy E_S would again be conserved. Entropy would be increased. However, order energy E_O

would go from -2 x 938 MeV to +2 x 938 MeV—from disorder to order. The order to disorder arrow would be reversed. This would be a reaction that would be prohibited by the second law of thermodynamics—unless the strong gravitational force that fuses the anti-semions would be stronger than the second law of thermodynamics (which otherwise governs weak interactions), which it is.

Here we see that the entropy arrow of time and the order to disorder arrow of time are separate and distinct, and are not one and the same thing. While all the reactions the author has studied increase entropy, the fusion of positron anti-semions reverse the order to disorder arrow, making more order out of the disorder.

Positron constituent electrino fusion might not only take the electrinos from disorder to order. It could make other physical processes in a local area go from disorder to order. The positron fusion not only violates the second law of thermodynamics, it reverses the order to disorder arrow of that law in a local area, making other processes in that area reverse. Let us consider that process more to see how it might be regulated.

We guess the desired relationships for reversing the order to disorder arrow in the second law of thermodynamics through dimensional analysis. We want to solve for r, the maximum radius in which the reversed law would be effective. There is a way we can obtain a length from combinations of our variables and constants. That way is in the right hand side of Eq. (10-2). The whole expression is the thermodynamic relation we are seeking. The thermodynamic relation is:

$$(\Delta E_o)_t > 0 \; where \; r < \frac{(\Delta E_o)_1 \; c}{ik}, \qquad (10\text{-}2)$$

where E_o is the order energy–the positive or negative energy in the pair production of particles; ΔE_o is the change in the order energy, where $(\Delta E_o)_t$ is the change in the total order energy of the system, and where $(\Delta E_o)_1$ is the change in the order energy for a single source reaction—for a positron fusion reaction it is approximately $2 \times 0.94 \times 10^9$ eV/collision $\times 1.6 \times 10^{-19}$ joules/eV = **3.0 x 10^{-10} joules/collision**; **c** is the speed of light—approximately **3.0 x 10^8 m/s**; we shall solve for the effective radius r; **i** is the effective beam collision current in each beam in Coulombs per second (we will solve for 10^{-11} or 10 picoAmps); **k** is the ratio of particle energy to particle charge. This energy per charge is the accelerated energy of the particle (0.94×10^9 ev times 1.6×10^{-19} joules/ev = 1.5×10^{-10} joules) divided by the charge of each positron (q = 1.6×10^{-19} coulombs), which equals **9.38 x 10^8 joules per coulomb**. The collision efficiency eff is not needed in this equation, because the result is not in particles, but is already in collisions.

Incredibly, the lower the current, the bigger the radius of the affected area. And the greater the current, the smaller the radius of the effected area. With 10^{-11} A equivalent collision beam currents, the effected radius r solves for 9.6 meters.

To get an idea of the positron collisions needed to reverse the order to disorder arrow of the second law of thermodynamics in what size of affected radius, see Table 2-1 and text on pages 13-15.

Remarkably enough, the affected area of second law reversal calculates to increase with the reduction of positron beam current. Area control is merely a matter of timed gating of the positrons in the positron-positron collider. [8]

6. Rate of Reversed Aging

The author will now calculate the rate at which reverse aging will occur in the calculable radius of the active Refresher: The beginning energy of the host particles (positrons) from which the fusion process takes place is $2m_ec^2$ per individual reaction. The ending energy of the host particles (protons) to which the fusion process tends is $2m_pc^2$ per individual reaction.

$$\frac{\Delta E_p}{\Delta E_{e^+}} = \frac{+2m_p c^2}{-2m_e c^2} \approx -1836.$$ This is a unit less expression from the available energy terms. What we seek is another unit less expression $\frac{\Delta t_r}{\Delta t}$, where t is the normal time during which a person or object ages, and t_r is the reverse time (negative) during which a person or object un-ages. The quotient is the relative rate of un-aging compared to aging. This also is a unit less quotient. What use of particle fusion parameters can yield such a unit less quotient? What terms are available to derive such a unit less quotient? What about the first terms and unit less quotient? If we equate them, we have $\frac{\Delta t_r}{\Delta t} \approx -1836.$ Reverse time would be negative and 1836 times as fast as forward aging time. Forward aging of 100 years would be un-aged in 19.89 days. Forward aging of 1 year would be un-aged in approximately 4.77 hours of machine time.

7. Miracle Working Power of the Refresher 1

The theoretical discovery of the order to disorder arrow in the second law of thermodynamics reverser (Refresher 1 for short) was a surprising turn, and engrossed the author for several years. By simply reversing the natural arrows between ordered events, many miraculous results were found to take place in theory.

What does it mean that the order to disorder arrow in the second law of thermodynamics is reversed? Events naturally come in order indicated by the arrows:

Healthy young adult→aging→wrinkles→aging→cancer→death→ cremation→scattering ashes

Reversing the order to disorder arrow in the second law of thermodynamics means all the arrows between the ordered events are turned around. The old and diseased become young and healthy. The clock is not really reversed. Adults do not become children again and disappear to extinction. The system just tends to maximum order, which is at young adulthood. Children still grow up to maximum order at young adulthood.

Many similar reversals can occur in the animal kingdom and the environment. The author imagined many marvelous things, but virtually forgot about the EFP Reactor.

8.   EFP Reactor in the Field of the Refresher 1

Finally the thought came, "What would occur if the EFP Reactor were in the field of a Refresher? The concepts of the effects assembled slowly. The accelerator electronics would not have resistive heating in the field. As a result the accelerator would be room temperature superconductive. There would not be any need for cryogenic energy losses. The accelerator would be about 100% efficient.

Reversing the order to disorder arrow in the second law of thermodynamics greatly affects all things with which we are familiar. But what would it do photovoltaic cells in a high energy gamma field? Outside the Refresher field, photovoltaic cells in the high energy gamma field would become damaged. They would become more and more damaged with time. This is a form of aging. What would happen if the aged photovoltaic cells were put in an order reversed Refresher field? The cells would un-age back to the original condition. What would happen if photovoltaic cells in an order reversing Refresher field were exposed to high level gamma radiation? They would not become damaged or aged. What would happen to the power that would ordinarily be absorbed in the aging process? Would it not be added to the power converted from radiation to electricity in the photovoltaic cells?

But what about the miscellaneous heating that would occur to photovoltaic cells in a high level radiation field outside an order reversing field of a Refresher? The heating process, though not necessarily damaging and aging, also occurs as an ordered process in the second law of thermodynamics. If the order to disorder reversed field of the Refresher were added, the photovoltaic cells would be cooled down. Heating would not occur in the field. What would happen to the power ordinarily lost to heating? Would not it be added to the power converted from radiation to electricity in the photovoltaic cells?

But what about the gamma photons that would not age the photovoltaic cells or heat them, but would pass through them without affecting them? What if the Refresher field were added, what would then take place? The next question can resolve this question. Is the shielding loss included in the order to disorder arrows in the reaction equations? Yes. Then with the addition of the Refresher field, the elusive photons would return or never penetrate the photovoltaic cells. What would happen to that power? Would not it be added to the power converted from radiation to electricity in the photovoltaic cells? This result is the hardest to take. We need experiment to settle this. If this paragraph were not true, we would expect it would take layers upon layers—many feet of photovoltaic cells piled on top of each other to stop the gamma photons. But if this paragraph is true, then gamma rays as well as sunlight could be stopped by a single layer of photovoltaic cells in the order to disorder in the second law of thermodynamics reverser of the Refresher. In the reversed field, the photovoltaic cells should be 100% efficient. An EFP Reactor must be built and operate in the field of a Refresher.

While an individual photovoltaic cell may be 100% efficient, it would not be possible to cover every spot around the reactor with photovoltaic cells. But it should be possible to achieve 60% to on the order of 100% efficiency—enough for the source to be self-sustaining and an energy source.

9. What about Radioactive Wastes?

As we now experience the second law of thermodynamics, neutrons + products → neutron activation products. Reverse that and activation products become deactivated and neutrons are given off. Another reaction involving neutrons: $n \rightarrow p + e + $ anti v_e. Reverse that and neutrons are produced. In the field of the Refresher 1, neutrons appear stable. Also in the field, radioisotopes are all backed out of existence. As long as the Refresher 1 field is on, the EFP Reactor will be radioactive waste free. Each annihilation of a proton or neutron on the inside surface of the collision chamber of accelerator leaves behind a new radioisotope. But with the Refresher, these do not decay with decay heat. Instead, these have electron capture and alpha particle capture and un-decay cooling.

References

[1] SUMMARY TABLES OF PARTICLE PROPERTIES, January 1, 1998, Particle Data Group, as quoted by *CRC Handbook of Chemistry and Physics, 80th Edition* (Boca Raton: CRC Press, 1999), pp. **11**-1 to **11**-49.

[2] SDI: technology, survivability, and software (Diane Publishing Co., May, 1988), p. 140, NTIS order #PB88-236245.

[3] Mathias Deckers, Steam Turbine Blading Technology for Siemens, Germany, "CFX AIDS DESIGN OF WORLD'S MOST EFFICIENT STEAM TURBINE," http://www.ansys.com/assets/testimonials/siemens.pdf.

[4] Stephen Hawking, *A Brief History of Time*—From the Big Bang to Black Holes (New York: Bantam Books, 1988), pp. 144, 145.

[5] Robert B. Leighton, *Principles of Modern Physics* (New York: McGraw-Hill Book Company, Inc, 1959), p. 665.

[6] Francis Halzen, Alan D. Martin, *Quarks and Leptons* (New York: John Wiley & Sons, 1984), p. 107.

[7] David S. Saxon, *Elementary Quantum Mechanics* (San Francisco: Holden-Day, 1968), p. 386.

[8] Gordon L. Ziegler, *Electrino Physics* (Lacey, Washington: Electrino Energy, 2010), Chapter 16, http://.benevolententerprises.org/, A much better up-to-date *Electrino Physics* Draft 2 is available at createspace.com, amazon.com, barnsandnoble.com and orderable at 2,500 book stores.

Chapter 11

Project Description

A. Introduction

The Project Description is the micro-designing, construction, testing, and operating in ever expanding footprint coverage of
the Refresher on earth and then on the sun, as the public demands
and the governments permit; and the mass production and prolifer-
ating of the Clean Energy Sources around the world rapidly. Our
objectives for this equipment are to make electricity free around
the world, to abolish sickness, pain and death around the world
quickly, to remove entirely all medical costs and funeral costs from
the people by treating them and healing them for not one cent in
fees; and retire the national debt and all debts public and private.
Other objectives are to halt and reverse global warming and to
tame the sun by backing it up several thousand years through the
Refresher technology.

B. Sketches

When the micro-design phase is done, we will have
complete blueprints for everything for the Refresher as well as the
Clean Energy Source. The only illustrations we have so far are
crude sketches:

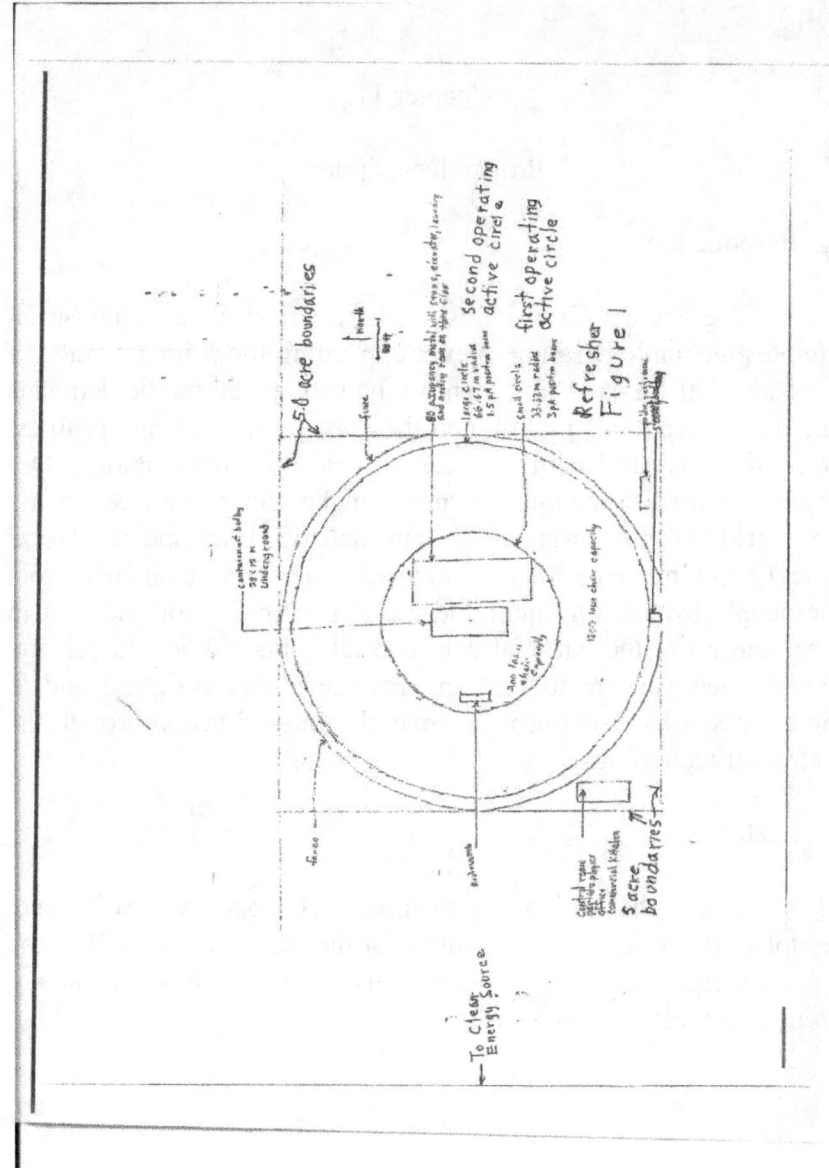

Refresher
Figure 1

5.0 acre boundaries

Second operating
active circle

first operating
active circle

5 acre
boundaries

To Clean
Energy Source

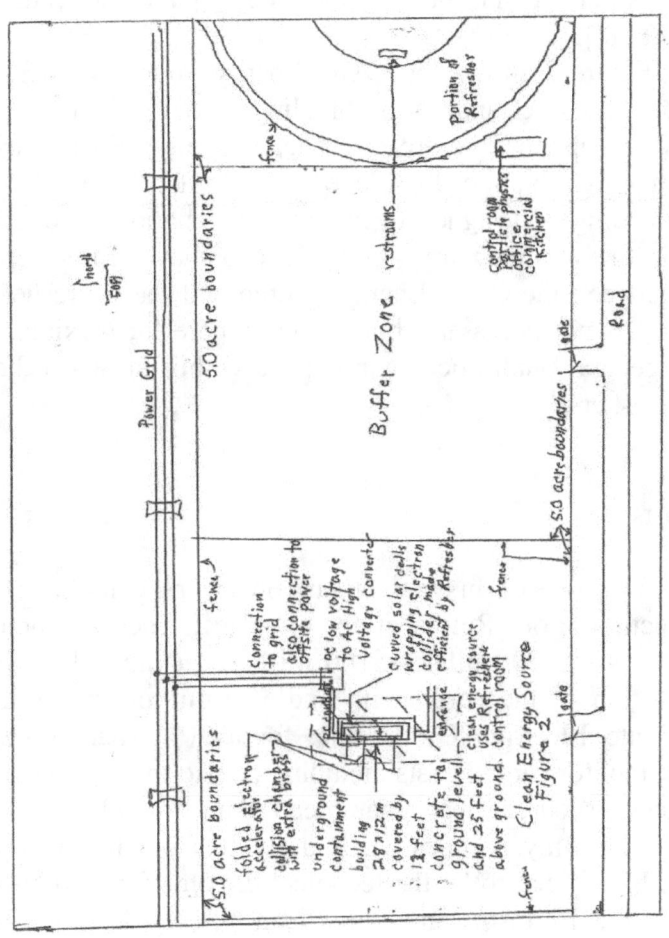

The figure contains the following handwritten labels (figure rotated):

North / Top

Power Grid

5.0 acre boundaries

5.0 acre boundaries

Buffer Zone   restrooms

Portion of Refresher

control room
particle physics
office,
commercial
kitchen

Road

5.0 acre boundaries   gate

fence

fence

fence

fence

gate

Connection to grid
also connection to offsite power

ac low voltage to AC High Voltage Converter

curved solar cells wrapping electron collider, made efficient by Refresher

clean energy source uses Refresher
control room

folded electron accelerator
collision chamber with extra boost

underground containment building
28ft x 12m
covered by 12 feet concrete to ground level
and 25 feet above ground.

5.0 acre boundaries

Clean Energy Source
Figure 2

Figure 1 is a crude sketch of the Refresher 1 showing also the placement of lodging and restroom facilities and control

73

building and security building. Circles in this figure show the perimeter fence and concentric footprint boundaries for different net equivalent collision currents [To obtain the net equivalent collisions per second, multiply the listed equivalent beam currents by 6.25E18.].

Figure 2 shows the Clean Energy Source with 37 feet of concrete and steel and earth shielding in the off chance that the Refresher trips off before the Clean Energy Source trips off, causing a great burst of high level radiation from the EFP Reactor. The shielding is designed to reduce the radiation to less than 2.0 mr/hr at the site boundary. Figure 2 also shows a five acre buffer zone between the Clean Energy Source and the Refresher, which probably is not necessary, but is shown here for maximum safety to the general public occupying space within the fence boundary for Refresher 1.

C. Costs

The costs of this project vary on several conditions. For the construction of one Refresher and one Clean Energy Source in one year it costs $130 million--$100 million guaranteed loan for the construction of the equipment, and $30 million investment debt equity into Electrino Group, Inc. for a 20% share in Electrino Group, Inc. for capital costs including fees to the U.S. Treasury for the senior secured debt guaranteed loan for $100 million for construction only. James M. Potter, Ph.D.—a linear accelerator expert, has agreed to be the technical director for the construction of both the Refresher and the Clean Energy Source (Electrino Fusion Power Reactor [EFP Reactor]) for one year for $100 million.

Costs

Construction costs for EFP Reactor and Refresher (guaranteed loan)
To James M. Potter Ph. D. JP Accelerator Works, Inc.

$100,000,000

Costs covered by debt equity by DOE or other sponsor $30,000,000

The $130 million above could be from an entity instead of DOE Loan Programs Office if they never finance the Refresher and Clean Energy Source. Or the $30 million can be an investment of debt equity by an entity facilitating a $100 million guaranteed loan by DOE LPO.

D. Emergency Situation.

The earth is going through a global warming crisis caused only in part by increasing greenhouse gasses caused by coal fired power plants and petroleum burning vehicles and homes. The far greater problem that effects global warming is that the sun is starting a nova sequence of getting brighter and hotter and hotter to peak at seven times as bright as normal for a day. The moon then will be as bright as the sun is now. If the sequence continues, the sun will go out in darkness, freezing the earth and its inhabitants! This does not have to happen! The Refresher machine that God the Father has inspired me with could actually reverse that nova sequence on the sun as well as regenerate the earth. But it has to be micro-designed first and constructed first. For $130 million, the Refresher and a Clean Energy Source could be designed and built in one year!